SONGS FROM UNSUNG WORLDS

SONGS

FROM UNSUNG WORLDS

Science in Poetry

EDITED BY BONNIE BILYEU GORDON

Science 85

BIRKHÄUSER   Boston · Basel · Stuttgart

Copyright © 1985 the American Association for the Advancement of Science

Library of Congress Cataloging in Publication Data
Main entry under title:

Songs from unsung worlds.

Includes index.
1. Science—Poetry.   2. Poetry, Modern—20th
century.   I. Gordon, Bonnie.
PN6110.S37S65   1985   808.81'9356   84-23677
ISBN 0-8176-3236-0

INDEXED IN

Grangers 8

CIP-Kurztitelaufnahme der Deutschen Bibliothek

Songs from unsung worlds: science in poetry / ed. by
Bonnie Gordon.—Basel; Boston; Stuttgart:
Birkhäuser, 1985.
   Aus: Science; 1985
   ISBN 3-7643-3236-0

NE: Gordon, Bonnie [Hrag.]

© The American Association for the Advancement of Science
ABCDEFGHIJ
ISBN 0-8176-3236-0 (0-8176-3296-4 paperback)
ISBN 3-7643-3236-0 (3-7643-3296-4 paperback)
Printed in the United States of America

Illustrations by Glenna Lang

CONTENTS

FOREWORD

"As poets over the centuries concentrated on Grecian urns, nightingales, ravens and romantic love, I am certain that poets in the future will focus on the configuration of planets, stars, weightlessness, and the discovery of our universe," says Maya Angelou, author of *I Know Why the Caged Bird Sings.* Only recently has science become a topic of widespread public interest, resulting in a flurry of newspaper and magazine coverage, inspiring television specials and movies. As science has inspired popular media and a new audience, it has inspired poets.

This anthology, *Songs from an Unsung World,* grew out of the successful reception of poetry published in *Science 85,* a popular magazine sponsored by the American Association for the Advancement of Science. During the last four years, the magazine has published science-related poems, and the editors grew to believe that a collection of them would appeal to many people. The project was proposed to Birkhäuser Boston, whose editors happily agreed.

The original concept was a collection that traced the influence of science on poetry over several hundred years. As I began to gather material for the book it became evident that Ms. Angelou's prediction was beginning to prove out already: There is such a wealth of science poetry being written today that an anthology based on contemporary work was demanded. The few historical poems included here were chosen because of particular relevance to one of the four themes by which the book is organized.

This anthology is organized in a somewhat unorthodox manner, by subject matter. The poems are presented in four broad categories: Poems about science and scientists, poems that record significant observations of the natural world, poems that use scientific language and metaphor to talk about love, death, friendship, etc., and, finally, satirical poems and poems that warn of the dangers of science and technology. Biographical information about the poets concludes the book, identifying many widely known modern poets and a substantial number of scientists. The living poets were asked why they use science in their work. The scientists were asked why they write poetry. Many of the replies appear in the biographical notes.

I have had a great deal of support and advice from Allen Hammond, *Science 85* Editor-in-Chief, and other magazine staff members, including Perry Turner, Julia Howard and Lynn Crawford. Generous help was offered from poet Karen Sagstetter and science writer Bruce Hathaway. And I received gracious assistance from poet George Starbuck and physicist Alan Lightman, who were called on to discuss poetry and science in a dialogue. That conversation was taped and is included in this volume as an introduction to the poems. Mr. Lightman, who teaches at Harvard University, is also a poet and a regular columnist for *Science 84*. Mr. Starbuck, who was trained in mathematics, teaches literature and writing at Boston University. Both writers helped me track down poems that appear here, as did my Birkhäuser editor Angela von der Lippe.

BONNIE BILYEU GORDON

INTRODUCTION

Dialogue between Alan Lightman and George Starbuck

AL / I think it's fascinating that we are considering poetry and science in the same book. The intersection of the two represents the full dimension of the human psyche: on the one hand needing to come to grips with the universe by quantifying it and rationalizing, on the other needing at the same time to express our unquantitative awe. That full, rich dimensionality is part of all of us, and there are some people who express one and some the other, and most of us express a little of both. It's also interesting that science, or mathematics—which is the language of science—is the most translatable of all languages, while poetry is the least translatable of forms, because the very slightest coloration of a word in poetry is important to the meaning of the poem.

GS / Mathematics is always trying to purge itself of accidental multiplicity of meaning, and poetry of course lives on that. The folksy, idiomatic text of a phrase will play against its literal semantics.

AL / So here we have poetry about science, bringing together these two dimensions—the untranslatable with the universally translatable.

GS / Some scientists I've known have always been the best punsters. Painters are, also. That's curious. When a real amateur at the business feels free with his writing he can sometimes do so much more. Science is a fairly new vocation, a new occupation for people to pursue. Now we have wonderful poets of other vocations who actually write about what they do and their worries about what it is they do. George Herbert, the English divine and metaphysical poet of the early seventeenth century, is a gorgeous example of somebody who wrote poems out of all the aspects of being a priest, worrying about whether he had a vocation, this kind of thing. Obviously there's going to be serious straightforward personal poetry about doing science. This is 1985 and we're in this jazzy century. Some of those guys won't sound at all like Herbert because they'll be writing about the fast lane, fast track. I mean Carl Sagan jetting from coast to coast—speaking at this symposium, the pop media, that kind of life. I also just look for the surges of excess creativity and emotion. There aren't going to be any people of Robert Oppenheimer's generation anymore, who write poetry (as he seemed to) partly because he wanted to be a sort of mandarin elitist. He was proving something to the other scientists. He was a more total man, a more Renaissance man than they were. Really, so much work was there, so relatively few of them, and there were so many late breaking things in science to do. Now, I imagine, there would be more crowding up in some of the fields. Scientists may not get as much satisfaction easily out of doing science.

AL / There are different ways in which scientists might begin to write poetry. They will be analogous to poets who are now referencing science. Some poets are using only the vocabulary of science, but are addressing the same things that poets have always written about, the basic human issues. Now that our lives are getting more and more affected by science, there's an increasing public awareness of science. And with the explosion of fast communications—particularly television—and more scientists writing about science, the scientific vocabulary has become a part of our culture, so, naturally, that vocabulary is entering poems. Poets also use science and scientific events as metaphor; again with the main subject matter being the things that poets have always written about. What I like best of the science poetry is written by poets who really understand some of the subject matter of science, the concepts, the way of thinking and the methodology, and the whole rationalist approach, using those things in poetry. In this case, the actual subject matter may not be scientific, the vocabulary may not be scientific, but the poem might be very closely identified with science and its analytic and quantitative approach to the world.

My analysis here is not as authoritative as yours would be, by a long shot, but I feel some of T.S. Eliot's poetry fits this category. I think "The Love Song of J. Alfred Prufrock" demonstrates the scientific way of thinking, though it doesn't deal with the subject of science.

GS / I have a cranky opinion, which is that Eliot represents something sort of counter-scientific. And yet you persuaded me at dinner last night that he represents a clarity of system of thought.

AL / I think the poem uses an analytical and rationalist approach. Eliot carefully takes apart the universe piece by piece to see where one moment fits in with everything else. And there is something logical and compellingly methodical, almost burdensome, about the way he gets to that point. It reminds me of the linear sequence of logical thought and analysis found in much of the scientific method.

GS / Well, I think that early in his career Eliot may have had in mind some very definite and prescribed religious methods, of self-questioning and meditation, various devotions, and this was a quality that he obviously admired in himself, but felt could be ridiculous, and in projecting a character like Prufrock he is intensely conscious of the danger of appearing ridiculous. What he's dramatizing there is close to some certain kinds of scientific method, clarifying and dividing the question, things like that.

L / In extreme form, the Euclidean method of proof in geometry.

S / Well, it is true, when you begin to examine the definitions, that poets use science, or elements of science, in widely varying ways. I'm really interested in how scientists write poetry. I'm hoping that readers will find this project fascinating and exasperating and utterly incomplete because they'll be looking for what must be all around us, what must be emerging—the poems of scientists. I too am most eagerly awaiting an input from elsewhere, from others into a closed little world that I spend too much time in, which is writers and literary scholars thinking about writing. What do you expect the scientists to be doing and saying and singing to themselves?

L / It's conceivable that many scientists' poems may not have any direct bearing on science whatsoever, but may rather provide alternative ways for scientists to express themselves.

S / One thing I wish I were going to live long enough to see in clear retrospect will be the love poems of scientists. Because some of the poets have given interesting intimations of how a radically altered, somewhat descriptive ecologist's perspective on a love poem can produce a touching but very different thing, like Auden's "Lullaby," which has this almost icy language in it that excited a whole generation. And it's easy to look back and see Dante and Petrarch as men steeped in the science of their time, such as it was. It's absolutely fascinating to see them moving at ease through their love poems, intent on showing them to other educated gents. And who will share these understandings and, as you say, the wonderful metaphors out of that cosmology? Poets will always be doing that. And scientist-poets will have to do that because the powerful metaphors tend to be reachings-out to what you feel or think you know of the powers around you in the universe. The poems I'm seeing remind me of an old love verse tradition, a very well established tradition, in a sort of patronizing way, the love song of an apothecary, or the passionate shepherd to his love. There's the whole pastoral tradition—gents being sort of patronizing about agricultural types. In fact if you look back through volumes of dusty old verse you see the passionate "blank to his love" is a common title and you're sort of making fun—for the shared amusement of some sophisticated audience—maybe it's the passionate astronomer to his love or the passionate biochemist to his love. The implication is that the fun of the poem is in the litigations of jargon you print. And of course some of the funniest and nicest and most pleasing poems of that genre have been written by the veterinarian or the apothecary. You find that washing over in poets like Donne and Marvell into serious poems which just sort of make a fleeting pretense—Donne will fleetingly attempt to be a cartographer—and have fun that way.

L / This would be using science as just one of another disciplines from whose perspective to view the rest of the world.

S / Yes, but partly just having fun, sort of taking the old subjects explicitly and deliberately and seeing what happens. And it seems to me to be a step beyond, in a direction that I really want, when I get Holub's poem "Evening in a Lab." I am very convinced that it's written out of a knowledge of what it is to be working late at night in a lab and to be discontent and to fantasize—actually he's fantasizing in the direction of ancient myth. But he's writing about a mood that seizes one at work. One of these poets, Alan Dugan, has a wonderful poem called "On Zero" where he starts off writing about the legendary, I think, nameless figure, the mathematician who invented zero. And so he starts:

> The man who first saw nothing
> drew a line around it
> shaped like a kiss or gasp
> or any of the lips' expressions during shock,
> and what had been interior
> welled from its human source
> and pooled, a mirror perilous.

I won't finish that first stanza, but he's starting off there not just playing around with a little bit of lore about mathematics, but a little lore of the anatomy and dynamics of facial expressions. Nobody knew how to think that kind of thing until this century—noticing that in extreme emotion, the face's repertoire is limited. But about eight lines further on he returns to his Arab mathematician:

> When his head, dead tired of its theory,
> dropped to the mark it made,
> his forehead drank the kiss of nothing.
> That was not sleep!

And he goes off into a dream, and I, who don't know, think that's a wonderful persuasive tribute to the pad-and-pencil scientist and the kind of replete exhaustion after you've thought of a great idea. And it's also a very funny thing where his head goes clunk like Charles Chaplin in a diner out onto his plate. So here's a poet playing around, showing off what he knows about science, but also giving what I take to be a touching, realistic view of how the scientific imagination might work.

I was wondering whether or not scientists sort of need other outlets occasionally for the irresponsible fantasizing they do—the cosmologist particularly. Maybe the last thing they'd want to publish is off-the-wall speculations about heat death of the universe ten to the thirty-first seconds from now or . . .

L / I don't know whether that sort of outlet or release of pent-up energy would take form within the scientists' poetic subject matter, or whether it would sort of dribble out metaphorically, lacking other themes. I suspect a scientist might lose a bit of his or her credibility by fantasizing on scientific subjects. It's much safer to fantasize on nonscientific subjects. Part of the definition of science requires an objectivity; that's how we scientists defend ourselves against the rest of the world, by substantiation and quantification. To reveal, in writing, your human side or your sloppy unquantitative awe at the universe, without shaking the whole structure, is tricky business.

S / I'm thinking more of the real necessary part of your equipment as a speculative theorist, you're actually trying out fliers for yourself . . .

L / All those fliers, to the extent that they have validity in science, are supposed to be backed by either physical arguments or quantitative calculations. It depends somewhat on what your definition of science is— but a thoughtful, twentieth-century philosopher of science, Karl Popper, said that science has to restrict itself to only those statements that in principle can be disproved.

S / That's using a razor's edge to define the rest as emptiness.

L / That doesn't mean that the rest is emptiness, or the rest isn't valid to raise; it just means that science has to limit itself in this way to get on with its business. A lot of these rambling speculations that you're talking about, if they couldn't be attached to some specific calculation or some experiment that could be done, they would be ruled out. Unfalsifiable speculations lie outside the framework of science and increase the speculator's vulnerability to attack. It might be much safer for the scientist to write about nonscientific subjects—imaginary subjects.

S / I think that's why so many scientists are consumers of mediocre, schlock science fiction. At least they were in my day. It's clearly far enough aside to allow speculation. But the scientist's mind needs that.

L / I think the scientist's mind does need that. I agree with you.

S / Implicitly we're thinking physics, astronomy, molecular chemistry, things like that, but if you get into developmental biology or genetics, then you find scientists who feel a little freer because the boundaries are clearer. Along the evolutionary line, for example, speculation about the origins of waterbird behavior and anatomy must be safe—like figuring when or why the bird developed an oil gland as sexual organ. There are ways that are acceptable within the literature for biologists to speculate decently about where that might have arisen and whether it derives from some sort of reptile behavior that pre-dated flight. I assume scientists speculate about things like that.

L / Up until recently there was a lot of controversy between those who believed in the punctuated equilibrium theory of evolution and gradualism theory. I think now there's a greater and greater data base, so there might be more consensus for the punctuated equilibrium theory. In those areas of science with less data base, there's more room for speculation. However, even the scientists in those fields must agree on certain ground rules, certain forms in which the speculation takes place. Many of them might be uncomfortable with sitting down and writing a pie-in-the-sky poem that wandered off in all directions from their subject.

S / I have some hope that science and the contemplation of science and the investigative ethic will give some poets a clearer sense of how they ought to act most of the time. As if there were an ethic of a proper productive behavior in the poetic imagination, a kind of imagination that flies off into fancy, building imaginary worlds and/or creating fanciful effects in clear but discreet forays. A sort of controlled creativity.

L / Creativity and expression may in fact motivate some scientists to write poetry, among doing other things outside their fields. I believe that creativity has more obstacles in science than in various art forms. Science alone may not be a sufficient outlet for the creative expression of some scientists, thus leading them to poetry as a way simply to express themselves. In science, there's such a large body of knowledge that one must tie into. I sometimes think of the difficulty of being creative in science—within the framework of science—as analogous to giving a painter just two colors and additionally requiring that he use only diagonal stripes, and then telling him to do a painting. A perfect illustration of this difference of creativity in science and in art is relativity and cubism. They were invented at about the same time, in the early 1900s. Today an artist has the choice of whether or not to include elements of cubism in his or her work, but a physicist developing a new theory must be consistent with relativity. I'm oversimplifying tremendously, but I do think there are differences in the ground rules and confinements. This difference could very well lead some frustrated scientists to trying their talents in other fields. There's a lot of pent-up creativity that's not being released.

S / Unhappiness is, after all, the great spur to becoming a poet.

L / There are intangible human questions that just don't fit into the province of science. And the opportunities for individual expression, even within science, are dwindling. Science is costing more and more money. As we snoop into territory that's further and further removed from human sense perception, we have to invent more and more elaborate and costly machines and experiments. Experimental scientists are unavoidably working in larger and larger teams. I believe this is going to affect the sociology and motivation of doing science in interesting ways. It may be that some scientists turn to other forms for individual achievement. The word "individual" is important here because when you write poetry or paint a picture, it's an intensely personal experience, and the essence of it is that individual expression. Whereas, in science the essence is the bottom line—after the work has been standardized and dry-cleaned and solidified. An experiment is valuable only if it can be reproduced; a theory is valuable only if it can be distilled into disembodied equations. The individual gets removed in that process and that is the power of science. But not much is offered here to people who want to leave behind something of themselves—to express their personal sonata.

S / That seems like a good, eloquent expression of one of the more serious motives. There's also another kind of poem that scientists might want to write, another sort of inspiration. Look at the rock musicians, at Bob Dylan for example. Look at Alfred, Lord Tennyson, and that is poetry as glee, in the possession of powers, possession of great brilliance and in fact it sort of shames our puritan ethics. You see a little bit of that in Sagan, whose manner as a rock star I don't much like, although I love what he's doing on the issues. Look at Freeman Dyson. That shy soul turns out to have a capacity for self-display after all. I'd love to see what poems he would produce, and he might. But the sense of being culture heroes—and some of the scientists surely are; they now have privileges and perogatives, they live grandly, they misbehave—that's something that we've always liked in our poets. Big bad examples, but good examples, too. It may be that they'll all end up doing crazy electronic music or something.

I don't know where the study of man is heading or in what mood, but the question "What is man?" has many more interesting answers now. We are clearly getting beyond centuries when science would come up with some possible new answer and the general cultural response was always, "Gasp, alas, yes man is but a mere beast," or "Man is but a mere machine," or "Man is a murderous predator." We have seen this response with recent things, such as territorialism. I suspect that more and more thinkers both in the humanities and in science are going to wake up to the notion that we're talking about man as we know him, this interesting species of thinking and worrying animal. Putting him in evolutionary perspective and putting him in cosmic space/time perspective doesn't change what we know about him. More tools for wonder. Understanding the mere magnitude of the circuitry in the brain and thinking about that doesn't make man an automaton. That's one of the things I'm most interested in watching. We may get over some of our panic. The culture has had a tendency to panic at scientific ideas. People will panic at entropy and then they'll panic at uncertainty or relativity—in each case they'll say "Ooh, that means 'God played dice with the universe.' "

L / What do you mean when you say that we'll get over these panics?

S / That we'll begin to see that these are all useful further perspectives to take on the human condition, on how we come to have our emotions and work them out. Each new discipline—sociobiology, for example—isn't offering an either-or choice.

L / I agree that all these developments have changed our perspective, but I don't see any indication myself that these sudden changes in perspective will cease. I think that new fundamental discoveries will continue taking place and we'll continue getting shaken up by these things, and I don't see that as an unhealthy situation. Following any major discovery there is an initial period, where it's painted in black and white, and then some years later we gradually begin to see the grays, and I think that will continue.

S / Well, T.S. Eliot felt that new Freudian and Darwinian perspectives on a secular and rationalist view of man presented an either-or crisis of belief. The only way that he could deal with this was to back off and turn his face away from the rationalists' future, and try to ingeniously, eloquently, find ways to turn back to the imaginary medieval, early Renaissance, village culture of Christian devotion. That kind of thing doesn't seem to follow for me. That's why Auden is an interesting contrasting case. He too was very serious about maintaining Christianity, being a believer. Yet every time he came on a new piece of scientific info or data, in a sort of sanguine optimistic way, but with genuine interest, he would show how this could be mapped upon. He would amuse himself and educate the rest of us by showing how much continuity and similarity could exist between ancient systems of thought and new ones. His "Ode to Gaea" is a wonderful little two-page anthology of geophysical and biological data and speculation—a satellite's-eye view of ocean currents and things like that, little tossed-off witticisms about pied pebbles that will soon be rookery islands. Here he managed, to my taste, always to find bearings between his respect for the intellectual traditions, his respect for Christianity and his consuming interest in what science was discovering without ever forcing it or faking it.

L / So you're using him as an example of someone who synthesized rather than created a panic, overturned the tables.

S / Yes, and he's very interesting as he looks at other people. Of course, his near-contemporary, William Empson, was a brilliant man and a brilliant literary critic and his poems had an almost unerring taste for the ephemeral and transitory and soon-to-be-demolished ideas. You can find all these mistaken geological ideas in the thirties, and Empson picks them up and makes brilliant little magpie poems which you read now and say "Yes, but it happened that was wrong." I don't know if science is willing to accept such a concept, but there is something like a taste in ideas, an inherent nose for what is central and important. You find Auden using the word ecology and talking about the ecology of Dante's inferno back in the late thirties . . . and in the last years of his life he wrote wonderfully about sensory deprivation experiments, working them into poems about that awful New York and the South Bronx, the rabid seething city.

L / I think both reactions to developments are useful.

S / You said you think to be upset, to be panicked, can be tonic.

L / I think so. Maybe the word panic is a little extreme, but I think being forced to totally rethink your view on a subject rather than search for all possible signs of continuity is healthy. I believe it's a healthy thing for science to get overturned every now and then, humbled in a sense. We need to keep in mind that all our theories—even ones that have been supported by experiment—are approximations to reality. On any given concept, there may be other ways of thinking. For example, a lot of apparent contradictions occur in quantum mechanics. The Heisenberg Uncertainty Principle for example—which states that objects cannot, even in principle, be pinpointed both in position and in velocity—is very hard for us to grasp because it violates our physical intuition of the real world. Matter appears to have concrete form, including the property of being localized. Many of our intuitive ideas about the world are limited by our basic sense experience: We're six feet in size, not one-billionth of a centimeter, and so on. We don't have a feeling for the hazy world of the atom. So we should be prepared to rethink the world every now and then, as our sensory limitations are peeled back one by one.

S / You take these things like the Heisenberg Uncertainty Principle, down at that scale, and any scientist who's involved in developing them or thinking about them realizes that. Those do not overturn probabilities at the tabletop level. What I'm talking about is panic reactions at the tabletop level, when people enjoy themselves by running around like Chicken Little asking why the sky is falling. I don't know where this tabletop is any more. But then again, people take refuge in flattering little halfway explanations which aren't true either. I've heard people say,

"Well, all we're talking about here is that we don't have any measurement down there, we can't get an instrument fine enough." Which is, I take it, not the case.

Well, in biology the world has been steadily overturned at the table-top level. People were of course immensely disturbed by Darwin and Huxley. To the point of rioting. English gentlemen losing their gentlemanly demeanor at Oxford. There was the Scopes trial. People also overreacted to the discovery of the powerful effects of hormones and psychoactive synapse activators. That our behavior, emotions, moods—things that people commonly associated with life crises and commitments of love and honor—can be produced on demand in a laboratory, introducing a few molecules into the blood; these kinds of things have seriously upset people. We could also go back to the circulation of the blood and what people thought about that discovery.

L / I gather that you would like to see some poets or maybe all poets and writers and interpreters of these things work in a way that maintains continuity with our full development without turning everything upside down.

S / Well, it sounds like I'm describing a sort of process which can boil down to stodgy conservatism and foot-dragging in the face of evidence. I just hope that people are willing to actively entertain whole new notions, turn their minds upside down, look at things in completely new ways, without feeling that this means that they have to abandon or deny their accumulated commonsensical sense of what's going on. I'd like to see people be much more interested in seeing the way a new perspective casts new light into the whole field of commonly apprehended experience.

L / What do you think would happen if tomorrow a spaceship landed from another solar system and some creatures walked out? How do you think that would affect our perspective?

S / Very differently than it would have twenty years ago, because we have all sorts of apostles out there trying to prepare us for this possibility. Let alone all the schlock fiction, *Close Encounters of the Third Kind*. It's easy to sneer at the drearily unimaginative, high-budget creations and versions of that. But yes, everybody, all scientists, like to think about that.

L / Do you think that would have an effect on our poetry?

S / How it happened would be all-important. It would be very unlikely that these creatures would be another species just at the threshold
of space travel as ours is. As Sagan and others say, there's more than a
99 percent chance that it would be creatures much more advanced or at
least more developed mentally than we are. So the big question would
be—after the sigh of relief that at least this takes the bottom out of arguments, this takes some moral problems out of our hands—whether we
were to be lab animals or meat animals. But little green men coming out
of a spaceship is grotesquely unlikely, in that it implies that there's been
no attempt at communications before. That implies typical sci-fi unlikely
ease of carting little creatures around.

L / I agree. I was using that ridiculous example to speculate—a very
dramatic development that would totally change our perspective. There
could well be things like that waiting around the corner, so that fifty
years from now, people might be writing about events and perspectives
that are totally unimaginable right now. Although, you're right, it proba-
bly wouldn't be a spaceship landing with little green men, it would be a
signal first.

S / Such a message could be very hopeful in the short term—like for
thousands of years—if people really believed it. But I think within a
short time people wouldn't really believe it, because there would be this
enforced long wait for the return message. I think that it's interesting
that we're getting into sci-fi stuff where everybody speculates, like
everybody speculates about the final atomic holocaust. The anthology
will have enough statements of that. That's the big, likely, obvious thing.
But, I feel very relieved these last few years, because I've found that scientists generally are best equipped to understand and face those very real
and irreversible dangers. The scientists have begun to group up and step
forth and take the lead in this. Five or eight years ago, one began to feel
a little spooky, because it seemed that the poets sounded like Chicken
Littles and weren't well enough informed. Ninety percent of the poetic
or novelistic attempts to alert people present these scenarios of catas-
trophe which were not very persuasive, just caper-movie jazz. So I'm
happy to hear scientists taking the leading roles. And I think these things
snowball. It's easier to see trends like this when the issues aren't quite so
central. It's really astonishing that since Earth Day—a decade and a half
ago—certain of the basic concepts of environmentalism are no longer

questioned anywhere except in a few right-wing pulpits out in the sticks. People have rapidly accepted the fact that ozone-layer depletion is a real thing and nobody is going around sneering at it, as if it were some kind of anti-fluoridation rot.

L / I'm wondering about poems that focus on natural history. Aspects of natural history, just like some aspects of astronomy, can get to be pure taxonomy. For me, that is not as interesting as the conceptual part of science. Just organizing a jumble of facts does not necessarily increase our understanding of the world. It's only when we can begin to synthesize those facts, to simplify the framework, that science achieves its power.

S / One of Amy Clampitt's poems uses natural history in a wonderful way. "Camoflauge" is about noticing a pebble-nesting, rock-nesting bird with eggs laid in the open. She finds this bird nesting in her asphalt driveway and doesn't just let it go at that, but makes wonderful remarks—ones nobody could have made thirty years ago—about the bird's alarm behavior, about the rising and crying in distraction. She briefly and economically alludes to the questions of how these behaviors developed. There's a way of being a botanizing, strolling naturalist. Richard Wilbur has some poems about birds in which he learnedly refers to some eighteenth- and nineteenth-century naturalists and how they described some of the diving birds who work the bottoms of streams. I think that there's room for poems like that here.

L / I think representations of those types of poems should be included, but we should distinguish between the various ways in which science enters poetry.

S / One way science enters poetry, which we've been alluding to, but not directly, is criticism of science and technology. So let's be clear. Also we need some poems that offer a clear, worrying criticism about the effects of a new level of information on the human species. Scientists have interesting ways to speculate on how immediately adaptable, how hugely adaptable we are. You can relate this to communications problems, that through TV and all kinds of other things, our brains are just stuffed. We don't know yet where the overload points are, but we see kinds of behavior that are typical of overload. Investigation of the danger points is in order. At the same time, we're in awe of our own iconic memory, millions of remembered scenes, little moments.

L / The rate of progress, of scientific and technological progress, is rapidly increasing. Every twenty years we're learning more than we did in all previous history. We're rapidly creating our own environment now; we've jumped off the slope of slow, gradual evolution. How are we going to interpret, how are we going to have time to analyze this onslaught of new information? And then distill it into poetry and art and other things? It will take some period of reflection to incorporate all of the material and understand what it means.

S / At its very best, one hopes that poetry has been a useful form of reducing and clarifying and simplifying, making a few chosen emblems stand for categories of experience and ideas. I'm not sure that this is a flattering description of poetry's function. But maybe this is one of the powers of poetry that will be more and more required and needed. We all want to believe that we can go out there and meddle with more things, and jack up society by its boot straps and get things going. This kind of speculation leads to the final, glum, conservative scenarios. There's either going to be a stultification or stiffening of society or some kind of imposed, or stupid, screening of information.

L / Or an uncontrollable specialization, where one person understands the wigwak, and another person understands the limlam machine, and few people can understand the whole picture.

S / OK. You've raised what I feel is the most hopeful megalomaniac, optimistic hope for this kind of collection. When scientists begin to write more poetry, more quintessential, carefully meditated things, it may be because they want to reach across those barriers of specialization and talk to each other. Because they do have a time, we all have a time talking to each other.

L / I suspect the question of increasing specialization is something we will have to face. It's going to be harder for poets to get on top of the whole thing.

S / Goethe could have done it, so could Petrarch. Auden got on top of a whole lot of it. No, but, you're right, it's going to be some Salvador Luria, or someone like that, who will next get on top of it. He won't come out of a damn writing program. When he steps out of his spaceship, I'll be willing to applaud.

How many scientists, just by your account are there? Not including the clinicians.

L / Oh, I would hate to try and estimate something like that. There might be a couple hundred thousand in the United States.

S / Approaching what you can get into the biggest ball park.

L / In this country.

S / That's clearly a factor of several hundred more than there were educated, potential humanists—sources for the population of writers in England up to the nineteenth century. Enough to produce Athens again. The golden age of poetry as a chosen avocation for some of the best minds. Because scientists are some of the brightest people. We steer them that way. The next brightest go into the law and disappear.

There is no chance that any book put together now, however conscientiously, will seem comprehensive and suitably representative and coherently ordered. But it's nice to think that this will be an intensely felt, if somewhat idiosyncratic, selection, including and involving science and science vocabulary, poets and scientists. I hope it will be not only refreshing for sporadic reading, but will have a lot of stimulating stuff in it. It may prompt some severely orderly minds to get furious and say "Yes, but also . . ." and "Yes, but there must be . . ." I do have this hope that it will remind scientists that some of the best furnishings of their own minds are here being travestied by interior decorators and poets. And it may force them to repossess some of this stuff and do it right.

SONGS FROM UNSUNG WORLDS

THE SUBJECT IS SCIENCE

Epistemology

I

Kick at the rock, Sam Johnson, break your bones:
But cloudy, cloudy is the stuff of stones.

II

We milk the cow of the world, and as we do
We whisper in her ear, "You are not true."

Richard Wilbur

Evening In A Lab

The white horse will not emerge from the lake
(of methyl green),
the flaming sheet will not appear
in the dark field condenser.
Pinned down by nine pounds of failure,
pinned down by half an inch of hope
sit and read,
sit as the quietest weaver
and weave and read,

where even verses break their necks,

when all the others have left.

Pinned down by eight barrels of failure,
pinned down by a quarter grain of hope,
sit as the quietest savage beast
and scratch and read.

The white horse will not emerge from the lake
(of methyl green),
the flaming sheet will not appear
in the dark field condenser.

Among cells and needles,
butts and dogs,
among stars,
there, where you wake,
there, where you go to sleep,
where it never was, never is, never mind—
search
and find.

                    Miroslav Holub

Žito the Magician

To amuse His Royal Majesty he will change water into wine.
Frogs into footmen. Beetles into bailiffs. And make a Minister
out of a rat. He bows, and daisies grow from his finger-tips.
And a talking bird sits on his shoulder.

There.

Think up something else, demands His Royal Majesty.
Think up a black star. So he thinks up a black star.
Think up dry water. So he thinks up dry water.
Think up a river bound with straw-bands. So he does.

There.

Then along comes a student and asks: Think up sine alpha
greater than one.

And Žito grows pale and sad: Terribly sorry. Sine is
between plus one and minus one. Nothing you can do about
   that.
And he leaves the great royal empire, quietly weaves his way
through the throng of courtiers, to his home
                                        in a nutshell.

                            Miroslav Holub

Graham Bell and the Photophone

"he yielded to . . . a fear of having moonlight
fall on him while he slept . . ." "he would . . .
if the moon were up, make his rounds to
shield his sleeping family from its cold,
uncanny light."

—*Robert V. Bruce,*
Bell: Alexander Graham Bell
And The Conquest Of Solitude

The undulating currents of his mind
early enabled Alexander Bell
to realize the telephone, dispel
the barricaded silences behind
the fastened portals of the deaf. Admired
as one "who fills the room where he appears"
and dreaming of a path to distant ears—
a voice astride a beam of light, unwired
by metal—why should Bell evade the grim
Diana's cryogenic fires? Above,
might she by photolytic art unwind
the restive undulations of his mind
or, with selenic stealth, intrude to hymn
subversive music to his sleeping love?

G. F. Montgomery

6

Hermann Ludwig Ferdinand Von Helmholtz

Hermann Helmholtz said the
  problem facing
the scientist is this:
reduce a creek, a kiss,
a flaming coal from this random
  tracing

to some irreducible final text
dancing to the air
of the inverse square,
and we are left with the question:
what next?

But there is always another layer
above, beyond, below
the last answer; we know
the scientist and poet shape their
  prayer

with Newton and Frost, who
  searched for order
instead of answers, and found
such grace in number and sound
they glorify the spell of light on
  water.

Peter Meinke

My Physics Teacher

He tried to convince us, but his billiard ball
Fell faster than his pingpong ball and thumped
To the floor first, in spite of Galileo.
The rainbows from his prism skidded off-screen
Before we could tell an infra from an ultra.
His hand-cranked generator refused to spit
Sparks and settled for smoke. The dangling pith
Ignored the attractions of his amber wand,
No matter how much static he rubbed and dubbed
From the seat of his pants, and the housebrick
He lowered into a tub of water weighed
(Eureka!) more than the overflow.

He believed in a World of Laws, where problems had answers,
Where tangible objects and intangible forces
Acting thereon could be lettered, numbered, and crammed
Through our tough skulls for lifetimes of homework.
But his only uncontestable demonstration
Came with our last class: he broke his chalk
On a formula, stooped to catch it, knocked his forehead
On the eraser-gutter, staggered slewfoot, and stuck
One foot forever into the wastebasket.

David Wagoner

8

Marie Curie Contemplating the Role of Women Scientists in the Glow of a Beaker

Self-luminous as her radium granules
Marie's thoughts
Took on new weight and number
That cast light impressions of possibility
Across the night-black templates
Of an open praire

Robert Frazier

Letter from Caroline Herschel (1750–1848)

William is away, and I am minding
the heavens. I have discovered
eight new comets and three nebulae
never before seen by man,
and I am preparing an Index to
Flamsteed's observations, together with
a catalogue of 560 stars omitted from
the British Catalogue, plus a list of errata
in that publication. William says

I have a way with numbers, so I handle
all the necessary reductions and
calculations. I also plan
every night's observation
schedule, for he says my intuition
helps me turn the telescope to discover
star cluster after star cluster.

I have helped him polish the mirrors
and lenses of our new telescope. It is
the largest in existence. Can you imagine
the thrill of turning it to some new
corner of the heavens to see
something never before seen
from earth? I actually like

that he is busy with the Royal society
and his club, for when I finish my other work
I can spend all night sweeping
the heavens.

Sometimes when I am alone
in the dark, and the universe reveals
yet another secret, I say the names
of my long, lost sisters, forgotten
in the books that record
our science —

Aganice of Thessaly,
Hyptia,
Hildegard,
Catherina Hevelius,
Maria Agnesi

— as if the stars themselves could

remember. Did you know that Hildegard
proposed a heliocentric universe
300 years before Copernicus? that she
wrote of universal gravitation 500 years
before Newton? But who would listen
to her? She was just a nun, a woman.
What is our age, if that age was dark?
As for my name, it will also be
forgotten, but I am not accused
of being a sorceress, like Aganice,
and the Christians do not threaten to
drag me to church, to murder me, like they did
Hyptia of Alexandria, the eloquent, young
woman who devised the instruments
used to accurately measure the position
and motion of

heavenly bodies.
However long we live, life is short, so I
work. And however important man becomes,
he is nothing compared to the stars.
There are secrets, dear sister, and it is
for us to reveal them. Your name, like mine,
is a song. Write soon,
                                    Caroline

                                              Siv Cedering

Letters from the Astronomers

If then the Astronomers, whereas they spie
A new-found Starre, their Opticks magnifie,
How brave are those, who with their Engine can
Bring man to heaven, and heaven again to man?

—John Donne

I   Nicholas Copernicus (1473–1543)

The sun is the center of the universe.
The planets move around the sun.
Yesterday when I went riding,
it began to snow. The seasons
change. The earth
turns
on its axis.

Share these ideas with your students,
but don't give them my name.
New continents are being discovered.
Books are being printed. Witches
burn. I am afraid
of the spirit
of the times.
What will they do, if I say
the sun is the center of this cosmic
temple? if I say
its distance from the earth
is infinitesimally small
compared to the distance
between the earth
and the stars?

I carry a pouch
of powdered unicorn's horn
and red sandal wood,
to cure
the ill. I have devised a new
monetary system.
I was asked to Rome to help create
a new calendar, but I declined.
My mathematics are not
sufficient. If you
have been trained in the art,
see if you can find the laws
that would prove
my ideas. Do you have a quadrant?
An armillary sphere? An astrolabe?

Siv Cedering

II Johannes Kepler (1571–1630)

*If we substitute the word 'force' for the word
'soul', we shall have the basic principle which
lies at the heart of my celestial physics.*

—Kepler

They say my mother is a witch.
She was arrested in the rectory.
They dragged her to prison in a trunk.
They want to put her on the rack.
For weeks she has been chained.
I am writing letters
asking them to release her.
My school has been closed.
The Protestant teachers have been burned
at the stake. My youngest child
died, of small pox.
But I try to continue my exploration
of a celestial science.
I have derived a musical scale
for each planet, from variations
in their daily motions around the sun.
A five-note scale for Jupiter.
Fourteen notes for Mercury, and Venus,
repeating her one long note.
Such harmony. As I picture each planet
floating within the geometric perfections
of space, I think geometry was implanted in man
along with the image of God.
Geometry indeed is God.

Siv Cedering

How Copernicus Stopped the Sun

"The fool will turn the whole science of Astronomy upside down. But, as Holy Writ declares, it was the Sun and not the Earth which Joshua commanded to stand still."

—*Martin Luther*

I like to picture it this way,
Call me what you will:

Nicolas looking out at the stars,
Their steady turn about Polaris,
Jupiter and Mars in retrograde,
Jupiter with only three bright moons
That night, and orange-red Mars,
Moving backwards across the sky.

He is dizzy from pickles
And harsh beer from Krakow,
Is saying poems under his breath
And swaying like a double star,
Doing sums and charting the sky
That will not stand or obey.

"Copernicus," he shouts, "confronts
The cosmos," and the cosmos staggers
And throws him to the reeling ground,
Where the stars spin round and round
In the night of the vernal equinox
With the whole sky upside down.

The rest is easy—how he understood
That nothing in this world stands still.
Or that black world of moons and stars
As well, how he knew that we curve
Around the sun in a six-ringed polka,
How we wobble into spring and fall.

He wrote it down at the very last,
The diagrams and circles, the *Revolutions,*
And they brought it to him on his bed
On the very last day, the bed revolving
And the room revolving and the day,
The day revolving, and the sun stood still.

R.H.W. Dillard

How Einstein Started It Up Again

"It should be possible to explain the laws of physics to a barmaid."

—*Ernest Rutherford*

"Oh, Albert," she said,
Shaking her head
And falling back
On the rumpled and unmade bed,
"You make everything
So dizzy." And he did.

It was Albert and his equations—
This equals that,
And it's all the same—
Knocked it all out of line,
Curved space, bent time,
Cosmos like a wrinkled sheet.

He made it all so energetic
And so odd, made attraction
Seem so natural,
Just like rolling off a log:
The up and down, the in and out,
First you're still, then you move about.

And the sun is all
In how you see it,
Boiling overhead,
Moving by, or the one
Still point
In the moving sky.

She found it hard,
How this could be that
And still be this,
How it's here and then there,
Or then there or then here,
Or then here and there, hit or miss.

She curved up like space
And then rolled in a ball,
Looking back
To just where she had been,
When Albert rolled in
And began to begin again.

Sun and still sky,
Red clouds that veer
And drift by, the day's demise
Or sunrise (depending
On where you are or were),
And Albert's continuing surprise.

"Oh, Albert," she said,
Shaking her head
And falling back again
On the rumpled and unmade bed,
"You make everything
So dizzy." And he did.

R.H.W. Dillard

The Observatory Ode

Harvard, June 1978

I

The Universe:
We'd like to understand,
But any piece, in the palm, gets out of hand,
Any stick, any stone,
— How mica burns! — or worse,
Any star we catch in pans of glass,
Sift to a twinkle the vast nuclear zone,
Lava-red, polar-blue,
Apple-gold (noon our childhood knew),
Colors that through the prism, like dawn through Gothic, pass,
Or in foundries sulk among grots and gnomes, in glare of zinc or brass.
Would Palomar's flashy cannon say? Would you,
Old hourglass, galaxy of sand,
You, the black hole where Newton likes to stand?

II

Once on this day,
Our Victorian renaissance-man,
Percival Lowell — having done Japan,
And soon to be seen
Doing over all heaven his way —
Spoke poems here. (These cheeks, a mite
Primped by the laurel leaves' symbolical green,
Should glow like the flustered beet
To scuff, in his mighty shoes, these feet.)
He walked high ground, each long cold Arizona night,
Grandeurs he'd jot: put folk on Mars, but guessed a planet right,
Scribbling dark sums and ciphers at white heat
For his Pluto, lost. Till — there it swam!
Swank, with his own P L for monogram.

III

Just down the way
The Observatory. And girls
Attending, with lint of starlight in their curls,
To lens, 'scope, rule.
Sewing bee, you could say:
They stitch high heaven together here,
Save scraps of the midnight sky. Compile, poll, pool.
One, matching star with star,
Learns that *how bright* can mean *how far.*
That widens the galaxies! Each spiraling chandelier
In three-dimensional glamour hangs; old flat nights disappear.
Desk-bound, they explore the immensities. Who are
These women that, dazed at dusk, arise?
— No Helen with so much heaven in her eyes.

IV

With what good night
Did the strange women leave?
What did the feverish planet-man achieve?
A myth for the sky:
All black. Then a haze of light,
A will-o'-the-wisp, hints *time* and *place.*
Whirling, the haze turned fireball, and let fly
Streamers of bright debris,
The makings of our land and sea.
Great rafts of matter crash, their turbulence a base
For furnaces of nuclear fire that blast out slag in space.
Primal pollution, dust and soot, hurl free
Lead, gold — all that. Heaven's gaudy trash.
This world — with our joy in June — is a drift of ash.

V

That fire in the sky
On the Glorious Fourth, come dark,
Acts "Birth of the Universe" out, in Playland Park.
Then a trace of ash
In the moon. Suppose we try
— Now only suppose — to catch in a jar
That palmful of dust, on bunsens burn till it flash,
Could we, from that gas aglow,
Construct the eventful world we know,
Or a toy of it, in the palm? Yet our world came so: we are
Debris of a curdled turbulence, and dust of a dying star
— The children of nuclear fall-out long ago.
No wonder if late world news agree
With Eve there's a creepy varmint in the Tree.

VI

The Universe:
. . . Such stuff as dreams are made on . . .
Yet stuff to thump, to call a spade a spade on.
No myth — Bantu,
Kurd, Urdu, Finnish, Erse —
Had for the heaven such hankering
As ours, that made new eyes for seeing true.
For seeing what we are:
Sun-bathers of a nuclear star,
Scuffing through curly quarks — mere fact a merry thing!
Then let's, with the girls and good P.L., sing carols in a ring!
Caution: combustible myth, though. Near and far
The core's aglow. No heat like this,
No heat like science and poetry when they kiss.

John Frederick Nims

The Concept of Force

Found myself seated by chance a few years ago,
at a luncheon for notables (I wasn't one),
next to Hans Bethe, the great German physicist.
I sat there, a layman in science,
trying to think of something to say,
and finally, after some awkwardness, managed a question:
"Dr. Bethe, in relativity theory — tell me,
has the concept of force between two bodies,
their mutual attraction, been superseded?
Do bodies behave as they do, coming together,
because space is as it is in the presence of bodies?
No force?" Dr. Bethe regarded me mildly —
he was a kind man, with a school-teacher smile.
"Force," he said. "The concept of force.
Has it been superseded? I think no.
I think the concept of force will continue to be —"
he paused, as if searching for words —
"effective and useful." That was all.

I give you out there, for your mulling over,
the words of a great physicist,
"The concept of force will continue to be
effective and useful." Since this is a poem,
some of you doubtless will think I mean people,
when I say bodies, coming together.
And that *that* is what will never be superseded.
And that Dr. Bethe was put in the poem
to assist in the metaphor.
That is not what I mean. There is no metaphor.
I mean only what Dr. Bethe said to me, smiling —
no more and no less —
about bodies in space and gravitational law.
Though of course you may think what you wish.

Robert Sargent

The Universe Is Closed and Has REMs

to Celia and Wally, to Milly and Gene

1.

One. one. one. one.
That's what God said.
Singular, singular, singular, singular,
Infinitely outspread.

Nothing under the nothing not even the sun.

No weight, no breadth, no negative, no north,
No three dimensions beckoning a fourth forth
With monumental pantomime, no buzz
Of energy-exchange, no instances.

It might have happened. Who knows which comes first,
The point flash, or some perfectly dispersed
Extrapolation into time-reversed
Of this explosion, this diaspora?

You like the God that freaks out and goes *"Unh!"*
I like the opposite extreme. How dull,
How uncollected, how bare-minimal
The necessary element might be.
One. one. one. Eternity.

Dead rudiments. None veer. None coalesce.

A pretty perpetuity unless —

Once one wants one twice two twice two the formalities take place.
Sub-elemental sarabandes objectify a space
Extravagantly sparse, and stilled, and stirless, but perhaps
Complicit. Something happening. Collapse.

2.

Absolute bash, and then

Sub-elemental smithereens again

But suddenly and picturesquely clotted
At every scale, down to the not-ness knotted
Into the nothing where a quark sleeps furious
And nags the smarmy noggins of the curious
With pointy-headed notions of the sphere
A quantum mass might shrink to, to cohere
Into its own black hole,
Inviolably sole
And satisfying to the theorist
Who likes his distillations with a twist
Of irreducibility at bottom.
Welcome to basic. Smoke 'em if you got 'em.
That does it for the first few trillion years.

Now then. If all you spacetime pioneers
'll dig into your briefing kits, behind
The replicative check sheet you should find
A rose-red pair of actuary googols
In case the credit mechanism boggles,
One windfall apple softened up with slug holes
To sniff at, and a parts list for these fogballs.
Mumble it to yourself if you get dizzy.
And hold tight. The next wiggle is a doozy.
Get used to it, 'cause you'll be getting lots o' these
Interchangeable pre-big-bang hypotheses.
And if the rose-red googols are a dud,
You bring it up at . . .

3. Question Period

Are we the first bounce, or the eightieth?

These dead immensities inventing death
Inventing difference inventing brilliance.
How many of them? Ten? A dozen? Trillions?

But if Xerxes
Had had Xeroxes . . .

If the aurochs
Got anthrax . . .

If the Mingel-Wurm
Returns . . .

No seriously. Look,
I mean I've read the book.
If having a trick thumb can tip the odds,
Why us-the-klutz? Why not cephalopods?
Put *brain* behind those graspings and completions,
Tune up those fine calcareous secretions
To gestate little lock-picks and escapements,
A quick squid'd run rings around these apemen's
Lathes and beams and hieroglyph prostheses
For busting big things down to byte-size pieces.
A thought could be the father to a stack
Of nacreous holography, played back
Instantly anywhere a clammy grip
Fondled the iridescent microchip.
And when it came to synthesizing sheer
Spectacle, to outface Poseidon's mere
Opulence and fecundity and scenery
With monuments of unaided Balanchinery —
Well — what I mean — why us?
Why not an offshoot of the octopus?

(If you knotted a cosmos.
If you twisted its wristwatch.
If you skidded it parsecs
In a picosec.)

But the Ik . . .
                          The euglena . . .
              The ozone . . .

4. Nap Time

Hush.

Everything in a minute. What's the rush?

It may work out. The big Let-there-be-light
May keep receding barely-out-of-sight.

Us prospectors can pan among the vestiges
With bright eyes and big dishes, nabbing hostages
But never quite contriving to decode
A backstairs access to the mother lode
That heaved up unimaginably once
And left this avalanche of evidence.
Awake, asleep, it streams right through our fingers.
Once in a blue moon, a neutrino lingers.
It drives these needle-in-the-haystack trackers
Crackers.

I like the latest inklings. Just this year
A theorist has posited a sheer
Propensity of Zilch to self-destruct.
That's what he says began it. That's what sucked
Somethingness out of Nothingness. Shazam.

ZILCH + ZILCH + NIL + NIL = NAUGHT

Cryptarithm in a puzzle book I bought.

I like the sense of *consequence.* Hot damn.

All this consequent middle. All this muddle.
Me the unlikely frog in so big a puddle
It puts the perfervid fancies of the priests
Back at about the level of a beast's
Diffuse imaginings of huger mangers
Where ever more companionable strangers
Scratch him behind the ears and pitch down food.

If only I could see how to preclude
Acting on every triggerable spin-off.

I can't. It's going to kill us, sure as gonif
And gizmo gravitate and groove.
The mills will grind; the merchandise will move.

I'm sorry Lennon died, but Lennon did.
What got him wasn't Belial-the-Kid.

Simple Possession loitered, and it pondered,
And found, until its concentration wandered,
a momentary sense of Rationale.

Recess time at the Okey-Doke Corral.

One possible response of the biota
Observable in front of the Dakota
Became the probable, since there was time.

Possession was nine-tenths of it, and I'm
Possessed. I have these Titans that I've paid for.
Polarises, Poseidons, Mark 12s made for
Me, and I haven't found a way to ditch 'em.

If I could tough it out like Robert Mitchum
And take the beating, confident the script
Would have me up, goof-balled and pistol-whipped
But standing, to choke back humiliation
And pick up the routine of my vocation
After the *big*-time baddies have their ball.

Not in the cards. We won't be here at all.
State-of-the-art has got way out ahead of us.

Dumb, then, for the merely not-yet-dead of us
To love the thing that kills us. But I do.

So beautiful, so various, so new.

Some times I want to bang their heads on the Universe and scream
"It's *beautiful,* you balmy bastards! THIS IS NOT A DREAM."

But no, I take my task as to record
At close hand, for the glory of no Lord,
Delight of no posterity, some part
Of what it was to take the world to heart
When all of it and more came flooding at us,
Absolutely positively gratis
And ravishing and perfectly disposed
To pal around with us an undisclosed
Number of million human generations
Until the Sun god goes on iron rations
And zaps us with a real survival crisis.

Wonderful: we just graduate from Isis
And Kali and Jehovah and all that
And start to see how hugely where-we're-at
Exceeds the psychedelic pipedreams of it,
And whammo.
            Tell the whole shebang I love it,
And buck the odds, and hope, and give it my
Borrowed scratched-up happy hello-goodbye.

George Starbuck

The Windy Planet

pole
polar   easterlies   pola
r easterlies polar easterlies polar easterli
es polar easterlies polar easterlies polar easter

roaring   forties   roaring   forties   roaring   forties   roarin
forties   westerlies   roaring   forties   roaring   forties   roar
ing   forties   roaring   forties   roaring   forties   roaring   forties

horse   latitudes   horse   latitudes   horse   latitudes   horse   latitudes

the trades the trades the trades the trades the trades the trades the
trades the northeast trades the trades the trades the trades the trades
the trades the trades the trades the trades the trades the trades the tr
ades the trades the trades the northeast trades the trades the trades t
doldrums doldrums doldrums doldrums doldrums doldrums doldrums doldr

the trades the trades the trades the trades the trades the trades the t
rades the southeast trades the trades the trades the trades the trades
the trades the trades the trades the trades the trades the trades the
trades the trades the trades the southeast trades the trades the trade

horse latitudes horse latitudes horse latitudes horse latitudes horse latitud

roaring   forties   roaring   forties   roaring   forties   roaring   fo
rties westerlies roaring forties roaring forties roaring forties roari
ng   forties   roaring   forties   roaring   forties   roaring   f

polar easterlies polar easterlies polar easterli
es polar easterlies polar easterlies pol
ar easterlies polar easterlies pol
pole

Annie Dillard

Orbiter 5 Shows
How Earth Looks from the Moon

There's a woman in the earth, sitting on
her heels. You see her from the back, in three-
quarter profile. She has a flowing pigtail. She's
holding something
in her right hand—some holy jug. Her left arm is thinner,
in a gesture like a dancer. She's the Indian Ocean. Asia is
light swirling up out of her vessel. Her pigtail points to Europe
and her dancer's arm is the Suez Canal. She is a woman
in a square kimono,
bare feet tucked beneath the tip of Africa. Her tail of long hair is
the Arabian Peninsula.   A woman in the earth.

                                                        A man in the moon.

                                              May Swenson

Note: The first telephoto of the whole earth, taken from above the moon by Lunar Orbiter 5, was printed in the *New York Times*, August 14, 1967. Poem title is the headline over the photo.

Space Shuttle

By all-star orchestra, they dine in space
in a long steel muscle so fast it floats,
in a light waltz they lie still as amber
watching Earth stir in her sleep beneath them.

They have brought along a plague
of small winged creatures, whose brains are tiny
as computer chips. Flight is the puzzle,
the shortest point between two times.

In zero gravity, their hearts will be light,
not three pounds of blood, dream and gristle.
When they were young men, the sky was a tree
whose cool branches they climbed,
sweaty in August, and now they are the sky
young boys imagine as invisible limbs.

On the console, a light summons them
to the moment, and they must choose
between the open-mouthed delirium in their cells,
the awe ballooning beyond the jetstream,
or husband all that is safe and tried.

They are good providers. Their eyes do not wander.
Their fingers do not pause at the prick
of a switch. Their mouths open for sounds
no words rush into. Answer the question
put at half-garble. Say again
how the cramped world turns, say again.

Diane Ackerman

Viking 1 on Mars—July 20, 1976

1

What throws
this shadow of
the left-handed fencer
across the bright clubfoot
laid on Mars?

Our faceless, earthly
trespass
done in a plume of exhaust:
engines off
as the first leg
touches down.
No pilot clambers from the tripod;
no Martian runs across the ground.

Sunlight casts
long afternoon shadows
straight as the earth-crow flies.
Makes the same
skeletal selves we know.
The Martian swordsman,
left arm extended,
right akimbo,
is invisible.
Only his dark, grotesque image
reflects on the burglar's footpad.

Without life of its own,
(high-born on the food chain)
Mars radiates
its mythic warrior-spirit.
Full-blown it springs from
red boulders,
rusty craters
and the light, thin air.
Territoriality
is the spirit's shield.
For a saber,
it accretes
the strength of legends.

Calls on Zeus,
the father,
and on the sun
to twist the earthly spaceship
into cargo-shadows,
menacing as a
consortium of suicides.

2

Shadows shortcut
the Lander's superstructure.
Like black rafts
shooting rapids,
they bounce over
the froth of instruments;
carry truncated angles of
legs and switches,
shock absorbers and gauges,
seismometer and ground scoop;
a lop of the outsize, foreign ear.

This flux of shadows
flutters the edges of
struts and nuts and bolts
until the dark lines compose.
Until the imminent adversary forms.
This is the Martian defender,
grounded on the earthly footpad,
his shadow weightless as computer signals.

3

Oblivious of danger,
its orders stored a year ago,
the Lander's camera aims
a nodding mirror.

Across two hundred million miles,
on ninety million screens,
earth people watch
the image of Mars build,
computer-line by line.
The way children wait for the picture
in the jigsaw puzzle,
the dot-drawn coloring book,
the decal transfer.

We shed our thin
slipper of maxims.
Better than stolen fire,
better than Olympian secrets,
we forget the forfeits
for hubris:
the whole soup of early lessons.

We risk our geo-center,
risk a new Original Sin.
We dare to look:
to see the arc of
red dunes, dust and boulders,
the brilliant sky of
another territory;
the ground unroll,
treeless, slow and sure
as a snake,
see it circle
the crewless freighter.

We know the Martian emblem,
already stamped on the Lander.
There will be no confrontation . . .
only the Seconds will fight the duel.

A new Bible begins.

Anne S. Perlman

The Cross Spider

The 1st Night
          A spider, put outside the world,
given the Hole of Space for her design,
herself a hub all hollow, having no weight,
tumbled counterclockwise, paralytically slow
into the Coalsack.
Free where no wind was, no floor, or wall,
afloat eccentric on immaculate black
she tossed a strand straight as light,
hoping to snag on perihelion, and invent
the Edge, the Corner and the Knot.
In an orbit's turn, in glint and floss
of the crossbeam, Arabella caught
the first extraterrestrial Fly
of Thought. She ate it, and the web.
The 2nd Night
          "Act as if no center exists,"
Arabella advised herself. Thus inverted
was deformed the labyrinth of grammar.
Angles melted, circles unravelled, ladders
lost their rungs and nothing clinched.
At which the pattern of chaos became plain.
She found on the second night her vertigo
so jelled she used it for a nail
to hang the first strand on.
Falling without let, and neither up nor down,
how could she fail?
No possible rim, no opposable middle,
geometry as yet unborn, as many nodes and navels
as wishes—or as few—could be spun.
Falling began the crazy web.

An experiment conducted in Skylab 2 in orbit around Earth in
1973 was to watch a Cross Spider, *araneida diodema*, spin a web
in space.

Dizziness completed it. A half-made, half-mad
asymmetric unnamable jumble, the New
became the Wen. On Witch it sit wirligiggly.
No other thing or Fly alive.
Afloat in the Black Whole, Arabella
crumple-died. Experiment frittered.

May Swenson

The Planets Line Up for a Demonstration

The occultist and the New York Times coffee man
read it over breakfast:
You are safest on high ground.
Watch for earthquakes, signs in the air.
Everything will be mutant and magic,
disaster, a beginning.
This is how we celebrate the universe,
a lot of people stuck in miracles.
We wanted prophecy and the refuge to watch it.
What everyone said was larger than the moment.

Whoever had a telescope or eyes knew
the high pitch of Mercury, heard Earth's
sad drone clicking into place like special beads,
marbles discovering order.
Wherever technology was, it wasn't far enough.
Voyager, somewhere sleeping in the distance.
The moon spread its silver weight
like opium on the diaphragm
lifting us higher than fear.

Josie Kearns

The Universe

What
    is it about,
        the universe,
        the universe about us stretching out?
We, within our brains,
        within it,
                think
    we must unspin
the laws that spin it.
            We think *why*
        because we think
            because.
        Because we think,
            we think
                the universe about us.

        But does it think,
            the universe?
                Then what about?
                        About us?
                If not,
must there be cause
            in the universe?
    Must it have laws?
                And what
            if the universe
                    is not about us?
            Then what?
            What
                is it about?
            And what
                    about *us?*

                                May Swenson

A Message from Space

Everything that happens is the message:
you read an event and be one and wait,
like breasting a wave, all the while knowing
by living, though not knowing how to live.

Or workers build an antenna—a dish
aimed at stars—and they themselves are its message,
crawling in and out, being worlds that loom,
dot-dash, and sirens, and sustaining beams.

And sometimes no one is calling but we turn up
eye and ear—suddenly we fall into
sound before it begins, the breathing
so still it waits there under the breath—

And then the green of leaves calls out, hills
where they wait or turn, clouds in their frenzied
stillness unfolding their careful words:
"Everything counts. The message is the world."

William Stafford

Jodrell Bank

Who were they, what lonely men
Imposed on the fact of night
The fiction of constellations
And made commensurable
The distances between
Themselves, their loves, and their doubt
Of governments and nations?
Who made the dark stable

When the light was not? Now
We receive the blind codes
Of spaces beyond the span
Of our myths, and a long dead star
May only echo how
There are no loves nor gods
Men can invent to explain
How lonely all men are.

Patric Dickinson

Ode to the Alien

Beast, I've known you
in all love's countries, in a baby's face
    knotted like walnut meat,
        in the crippled obbligato
    of a polio-stricken friend,
in my father's eyes
    pouchy as two marsupials,
        in the grizzly radiance
of a winter sunset, in my lover's arm
    veined like the blue-ridge mountains.
To me, you are beautiful
        until proven ugly.

Anyway, I'm no cosmic royalty
either: I'm a bastard of matter
    descended from countless rapes
        and invasions
    of cell upon cell upon cell.
I crawled out of slime;
    I swung through the jungles
        of Madagascar;
I drew wildebeest on the caves at Lascaux;
        I lived a grim life
    hunting peccary and maize
in some godforsaken mudhole in the veldt.

I may squeal
from the pointy terror of a wasp,
or shun the breezy rhetoric
of a fire;
but, whatever your form, gait, or healing,
you are no beast to me,
I who am less than a heart-flutter
from the brute,
I who have been beastly so long.
Like me, you are that pool
of quicksilver in the mist,
fluid, shimmery, fleeing, called life.

And life, full of pratfall and poise,
life where a bit of frost
one morning can turn barbed wire
into a string of stars,
life aromatic with red-hot pizzazz
drumming ha-cha-cha
through every blurt, nub, sag,
pang, twitch, war, bloom of it,
life as unlikely as a pelican, or a thunderclap,
life's our tour of duty
on our far-flung planets,
our cage, our dole, our reverie.

Have you arts?
Do waves dash over your brain
like tide rip along a rocky coast?
Does your moon slide
into the night's back pocket,
just full when it begins to wane,
so that all joy seems interim?
Are you flummoxed by that millpond,
deep within the atom, rippling out to every star?
Even if your blood is quarried,
I pray you well,
and hope my prayer your tonic.

I sit at my desk now
      like a tiny proprietor,
a cottage industry in every cell.
    Diversity is my middle name.
My blood runs laps;
    I doubt yours does,
        but we share an abstract fever
           called thought,
a common swelter of a sun.
So, Beast, pause a moment,
        you are welcome here.
    I am life, and life loves life.

Diane Ackerman

Footnote to Feynman*

Science takes away from the beauty of
    the stars?
On Earth, stuck on this carousel my
    little eye
(atoms: my stuff was belched from some
    forgotten star)
my eye can catch one-million-year-old
    light. Do I
see less or more? Mere globs of gas?
    Nothing is "mere."
I see them with the greater eye of
    Palomar —
What is the pattern, or the meaning, or
    the why?

Earth, stars, a vast pattern — of which I
    am part —
(and the whole universe is in a glass of
    wine)
stars rush apart from common starting
    point. My heart —
red as Betelgeux, Antares, Aldebaran —
my heart beats to the mystery of the
    sky.

It does not harm the mystery to know
    our birth:
The stars are made of the same atoms as
    the Earth.

*Adapted with permission

Jonathon V. Post

First Rainfall

On dry days, I remember
the first rainfall on earth.
Clean and undesigned,
my atoms were there.

Clutching at hydrogen, bloated
with ammonia clouds, the air
no longer held,
and gushed

its first relief.
The wind was wet in sheets,
each force moved in its own sound.
What sod there was turned moist.

The earth, still rounding
like a newborn's head,
gurgled in the fog
and freely outgassed.

Alan P. Lightman

The Island of Geological Time

for Stephen Jay Gould

On this island, species subdivide so fast
sometimes you have to say, "Look!
The feathered descendant of the fruit bat
just ate the curly-tongued hummingbird's
six-toed offspring!" Brontosauri
and Iguanodon became extinct last week;
now engineers from Exxon
are drilling for oil.

          My hotel
was inland when I checked in
but on the beach by sundown:
It's the local plate tectonics.
Whales can't stand these waters
where the sinking Earth outsings them,
though cellists from around the world
like to tune their strings to that
'E' below sea.

          The moon lies so low in the sky
it goes whiz-bang from horizon to horizon:
The ocean sizzles and whole schools of fishes
turn warm-blooded, crawl ashore,
and bear live young in the morning.
The constellations swerve and dodge
like hopscotch, and poor men bet on where a star
will rise next evening.

Laura Fargas

47

Brevard Fault

The crack in our hearth of land
runs almost imperceptible under hills
and riverbottoms, slicing ranges.
So much folding and drift and erosion
have covered the flaw
the terrain appears seamless
except for the belt of crushed rock
strewn across the drainages.
But the split, deeper than any spring or cellar,
older even than the rise of mountains,
touches the fire of the original
schism, and reaching beyond highland,
piedmont and coastal plain, divides
the continent against itself
and builds the grudge of kinship
under quiet blue slopes.

Robert Morgan

Time's Times Again

The fall of deepbottom Arctic water
down the Atlantic midrib, a glacial inch a
month, the high assimilations of the free-wandering

jet stream, these places we look to for durance
or dwelling motion — sometimes, ephemeral, we need
them so much (or the floating apart of continents,

a centimeter a year, into the lank voyages) we
forget the many shimmering little absolute
disappearances, goings-away like local problems

solved (hair in the drain pipe) whereas the big
problems dwell in unsolvability's sway, useful
as systems of lasting definition, too big to be

let go of, currencies: everything is saved in the
disappearances & returns except what we love, the
particular jet of light, like a remark, in the particular

eye, that is the construct whose fire is so nearly
inexpressible we think the thinnest, highest
meanders of ozone not so crushing: what, completely

away, however much is left where so much, lakes,
clouds flow without loss, oh, well, though lakes
and clouds can't keep either: and, of course, the

axis has shifted more than once and Arctic migrated
down and around and the jet stream, before oxygen,
bore a different tune: all is lost — there

is an ultimate mere brightness not much having to
do with our business, though the energy base of
any business: still, and specially since we stay

a blink or so, the brightness in eyes,
having to do at times with love or in one time
the understanding of final pain, I go with that.

A.R. Ammons

Unpredictable But Providential

(for Loren Eiseley)

Spring with its thrusting leaves and jargling birds is here again
to remind me again of the first real Event, the first
genuine Accident, of that Once when, once a tiny
corner of the cosmos had turned indulgent enough
to give it a sporting chance, some Original Substance,
immortal and self-sufficient, knowing only the blind
collision experience, had the sheer audacity
to become irritable, a Self requiring a World,
a Not-Self outside Itself from which to renew Itself,
with a new freedom, to grow, a new necessity, death.
Henceforth, for the animate, to last was to mean to change,
existing both for one's own sake and that of all others,
forever in jeopardy.
                The ponderous ice-dragons
performed their slow-motion ballet: continents cracked in half
and wobbled drunkenly over the waters: Gondwana
smashed head on into the under-belly of Asia.
But catastrophes only encouraged experiment.
As a rule, it was the fittest who perished, the mis-fits,
forced by failure to emigrate to unsettled niches, who
altered their structure and prospered. (Our own shrew-ancestor
was a Nobody, but still could take himself for granted,
with a poise our grandees will never acquire.)
                          Genetics
may explain shape, size and posture, but not why one physique
should be gifted to cogitate about cogitation,
divorcing Form from Matter, and fated to co-habit
on uneasy terms with its Image, dreading a double death,
a wisher, a maker of asymmetrical objects,
a linguist who is never at home in Nature's grammar.

Science, like Art, is fun, a playing with truths, and no game
should ever pretend to slay the heavy-lidded riddle,
What is the Good Life?
                    Common Sense warns me of course to buy
neither but, when I compare their rival Myths of Being,
bewigged Descartes looks more *outré* than the painted wizard.

W.H. Auden

Ice Dragons

In a museum we find them
where they fell:
ichthyosaurus
with seven dragon whelps
in her belly;
sail-backed stegosaurus,
an armor-plated goon
wielding ratchety paws
and eye-coddling breath.

A pinafore of scales,
the sauropod toddles,
fanning its tail
through the mud
as it vamps
from bayou to sandpile,
teeth big as loaves,
a rosebud for a brain.

Another dips
a gravyboat head to drink,
while bird-monsters
on shoe leather wings
snuff the quickness
from a shrew.
Squat lizards spit bile,
and baggy-throated tots
trot after prey
with pipette-like claws.

Did they live on to test
Galahad and St. George?
Did they feel
the sudden whammy
of a global gasp?
We blizzard guesses
at their habitat.
We puzzle who
or what's to blame.
Only the bare bones
of a life remain.

Diane Ackerman

Fossils

I come down across stones lightly,
a part of them. Limestone, shale,
something else that's old-bone white—
perhaps the granite knows.

(The translation of time from stone
   to stone
      takes time. Things
move slowly.)

Trilobites mix quietly with small fishes.
Coal knows more by far than I.
Anthracite blinks in the sun,
smiling sleepily, thinking deeply of seedferns.

There was a point there
when things fought to the death
to decide whether a clutch of eggs
would bear scales or feathers.

But now the echo of Archaeopteryx
is just a clumsy arrow bent in sandstone,
with a three- or four-chambered heart
that still sighs with your ear held close.

Arthur Stewart

The Origin of Species

1

Corncrakes

Not to be confused
with cornflakes,
the Corncrake
possesses long toes
adapted for swamps
though he prefers
meadows.

Like a woodpecker
on a beanstalk.

2

Cabbage, varieties of, crosses

The exotic cabbage-flower
is tricky, with its pistil
surrounded by bodyguards—
those six stamens—and its
self-supplying pollen.

Stick to mongrel cabbages:
they're less high-strung.

3

Beetles, wingless, in Madeira

Was it indolence,
this sitting around
leisurely under wind
lulls, hardly moving
for days, that finally
did it, first shortening
the feathery wing,
weakening the muscle
which ran its tiny motor,
then sending it forever
into disuse?

Lazy Beetle, scientists
call it, but Beetle
took his time getting here,
drifting for centuries
on an old log in the sea.
When he arrived, he'd
made up his mind: this
was his last stop, here
on the comforting sands
under the sun in Madeira
where the music was much
to his liking.

4

Butterflies, mimetic

The mockers and the mocked
always inhabit the same region.
We never find an imitator living remote
from the form which it imitates.

The mocker is almost invariably
a rare insect, whereas the mocked
abounds in swarms.

The mocked keeps the usual dress
of the group to which it belongs,
whereas the counterfeiters
change their clothing frequently.

5

Falconer, Dr., on naturalisation of plants in India

Dr. Falconer has taken it
upon himself to urge
the cardoon and a tall thistle
to take out citizenship
without delay.

Six months residency
and a short examination
plus the national anthem
and they will be full-fledged flowers
capable of voting, paying taxes
and complaining to their congressmen.

It will be some time, however,
before they will be permitted
to purify themselves in the Ganges.

6

Bentham, Mr., on British plants

How many—Gall-of-the-earth,
Miterwort, Solomon's-seal, Bee-balm—
have been falsely ranked.

Why, Mr. Babington once listed
251 species, whereas Mr. Bentham
gives only 112, leaving a difference
of 139 doubtful forms consigned
to standing for all of time
alongside the virtuous pagans,

those born prior to the light
of God. Oh Pagan-poets, wild flowers—
chums without category or portfolio.

7

Forms, lowly organized, long enduring

You would not ask, say, the intestinal
worm to change himself, grow an enormous
multilobed brain scrawled with wrinkles,
to remember perfectly each place he'd lived
and speak of it in clear sentences.

Yet he goes on, making the most
of his visitations, taking what little
comes his way. His requirements are small.
No one asks him to sign a check
or keep an eye on the children.
He stays confined to his peculiar station,
just out of harm's way, not too much
competition, no chance of a favorable
variation.

                              Myra Sklarew

Progress?

Sessile, unseeing,
the Plant is wholly content
with the Adjacent.

Mobilised, sighted,
the Beast can tell Here from There
and Now from Not-Yet.

Talkative, anxious,
Man can picture the Absent
and Non-Existent.

W.H. Auden

Peking Man, Raining

Littered around Peking, Choukoutien
fossils: insects, tigers, birds' wings.
Those small squalls fanned the grey rock,
wriggled among terrestrial kingdoms
for centuries.

They called them "dragon's bones"
and dug them up, ground them into potions
for fevers, medicinal teas. Just one dose
from the benevolent beast eased the soul's
unrest.

A wonder there were any left. Shown
a single tooth, Black wore it in a locket
for years and envisioned the missing link
as he picked the clay. He knew an ape-man's
molar when he saw one. Though everyone
laughed.

All he found were flaked pebble
tools: stones fractured along the ripple,
cut gracefully, scalloped from the natural
vein, quartzite or chert, flayed down
to fit the hand.

Listen: the body's suffusive, blends
with the Earth. But a chopper survives.
Dropped in the caves around Peking, those lost,
unburied seeds of ourselves left trails
like breadcrumbs.

He followed, stored them in his marrow
like heartbeats. Broken, abandoned, or blunted
by pounding, an abundance of implements, skills
and diversions were locked in the limestone.
Years later:

rooted in sand and travertine, the skull
of a man who stood behind us, mute, eyes shaded
to gaze at the moon or doze in the sun.

Small-brained, but upright. Human, like us,
he suffered and lay down, began to rain,
to flower.
   A slow leaching of bone back to the Earth.
Atom by atom replaced by stone, a rich mineral
infusion, imbued over time. Until ancestral
remnants besprinkled China like willows,
sinking in soil, fine as the lines in your palm
or branches of a tall tree deep in the middle
of a forest.

Katharine Auchincloss Lorr

Skull of a Neandertal

The teaching assistant hands it over
casually, like a rag or
repair manual.

The jaw gone.
The skull
walnut brown. A chunk is missing.

"Murder?" She shrugs and puts it
back into the drawer.
She pushes the drawer shut with
   her hip.

Outside a cyclist leans forward
shifting gears over

ginkgo leaves
the color of teeth.

Michael Cadnum

Hominization

Lucy,
blessed among women,
three million years ago,
when there were no legends,
just the loving search
for dandruff in fur,
was waiting for someone
on the shore of the lake,
no one came but death,
the appropriate death for the species
Australopithecus afarensis,
a four-foot death with a man's step
and a monkey's skull.

Lucy — excavated,
partly assumed into heaven,
after some filling with plaster,
the forgotten nursery rhyme,
eeny meeny miney moe —
is still waiting.

And so are we, maybe,
despite plaster shortages.
And so are we,
with Pliocene hopelessness,
on the shore of the lake.

And maybe they'll find us one day,
when people finally exist.

Miroslav Holub

Two Sonnets

Love Song to Lucy

Three million times your bones have swept around
The sun since last your warm brown foot walked here
Upon the veldt. The hills, the lake, were dear
To you, and morning flowers, and each sweet sound
Of bird. In meadows where you played were found
The beasts that fed and clothed you—life's career.
And so you lived, until one day in fear
You died, and never knew you would astound
A future race. What waves of time beyond
Your ken evolved your sires, and ours, and sent
You here to us upon this destined shore,
Where we, your seed, have found you and respond
In awe? You speak in tomes you never dreamt—
A parent-link to all that lies before.

Lucy Answers

Your turn will come—time upon time your bones
Will also sweep the sun, and from the clay
Strange creatures, on a far and stranger day,
With eye and hand the primal mind disowns,
Will find you there among the silvered stones—
Will lift you, brush the ancient years away
And sift your possibilities the way
You do with me, in hushed and puzzled tones.
Your seed will seek his sire in mark and line
And try to mold your face, as you mold mine.
Yet I knew not you'd issue forth from me,
Nor can you penetrate his mystery.
As silence holds all future time at bay,
So tides will turn and sweep him, too, away.

Helen Ehrlich

64

Jutaculla Rock

Up in Jackson County they have
this soapstone in a field, scored all over
with hieroglyphics no one has solved.
Or maybe it's more picture writing,
the figures so rough and worn
they no longer represent. But
the markings are at least distinct
enough to tell they're made with hands
and not just thaw and wind and water.
Among the written characters of all creation
that big rock seems significant, but
like the written characters of all
creation is unintelligible without
a key to its whorls and wisps
of scripture that seem to shiver
in the rain on its face
like fresh-inked chromosomes
or voice-prints of quasars, there
in the washed-out cornfield. But what
could be more awesome than a message
ancient, untranslatable, true,
up where giants walked on the balds,
in text and context, history, word?

Robert Morgan

Gate

Through ignorance we arrived late
at the exhaustively excavated,
partly restored palace of Knossos.

The site was still open
but crowded with tour groups
we had to sidestep in a rush,

eavesdropping in passing, if
possible, in case some guide said
a sentence which mattered, until

wardens began blowing mock
herdsmen's whistles to chase out all
foreigners like docile-stupid sheep.

Anyone who tries to make sense
of the fragmentary pavings & broken
monuments of his own speeding life

knows the sensation, as he is driven,
same as ancient courtiers & slaves,
towards a gate   past which   being
ceases, let us say, to surprise.

David McAleavey

The Question

All of us believe
we were born of a virgin
(for who can imagine

his parents copulating?),
and cases are known
of pregnant Virgins.

But the Question remains:
from where did Christ get
that extra chromosome?

W.H. Auden

Lovely Girls with Flounder on a Starry Night

This winter the ocean rushes west
parallel to the torn beach at Water Mill

& at Mecox it has washed through the inlet
translating the sand's glyphs
codes of an unformulated pale language

a lonely hybrid
its single lyric fragment
oscillation of a lost DNA strand.

The Atlantic, with its littoral drift, listens
melancholic under a mackerel sky
an homogenate, its seismic reflection profile
a Eucharist, a viscous template
its trace elements seeding
its amorphous & primeval molecular sieves

yellow-tail & fat black-back flounder
languid rafts in enzymatic mud.

From Purgatory Hollow to Towd & Cobb
lovely girls of this outwash plain

rapacious birds & grazing ungulates
lament this, the moon's probes.

They are the primitive food gatherers
the first human inhabitants of this region
the haunted spirals of their songs are protected

groves of angiosperms their gauntlet
root nodules, methanogens, denuded larvae
with flask-shaped hearts
their sustenance.

Even so, along the outer bar
the *Shinnecock* & *Pauline* still drag for fish
silent tribes, their music coagulated
in their parenchyma, never heard
by the seraphs of killer satellites

nuclei of these heavens, prowling
in a syncytium, calling to the girls

recombining their DNA while they sleep.

Whatever, this much we are sure of:

at a recent South Fork Consortium on Virology
& Continental Reflection Profiling
new scanning electronmicrographs
showed Mecox's membranes
to be singing their own music

unattended shorelines, equinox passages
isolated deprivation chambers
meridian transits, sphincters, shunts

harmonic coefficients & orbital elements
of tidal suction

causing a diatom bloom
nitrogen fixation

& winter flounder to inhale bait

quietly, giving no inkling
of their presence as they prowl
the paleoradius of polymers

the girl's waists

orifices of mammalian chromosomes

in chilly waters
of Mecox's salty coves
where a strong wind-shear wails.

This ancient land-bridge, Mecox
still a radius of gyration
in hollow senescence

it keeps turning away from me
a galactic recession

like those girls.

Anselm Parlatore

The Supremacy of Bacteria

The vast Pre-Cambrian microbial mats,
Exiled, enfossiled into stromatolites,
Barely survived four million millennia
Spread thin through soil and water.

Yet their necessity still haunts our intestines,
Bonding the vegetable in matrimony
To that noble gas nitrogen.
They practice a most sophisticated chemistry.

Do not discount their gypsy influence.
Bacteria may have engineered us solely
To explore lunar dust and Martian clays
For their most distant of cousins.

Robert Frazier

Cancer Research

The Tumor Virus workshop is finished
but the juice of a new protein moiety
stains my fingers. Outside the hall
an arid wind suffocates the rose bushes,
light filters through the lewd damasks
& velvets lining the windows of the street.
Black hooded baby carriages pass me
on my way to the bay shore
where viruses relax in neutral currents
& the membrane fluidity
of the estuarine invertebrates
becomes evolution's hinge.
The lipids & proteins vibrate
a scaffolding of their own
planar self-array
the clock's *in vivo* silence.

Anselm Parlatore

Neural Folds

for John Teton

The frog embryos spin,
a million tiny skaters
in bright sacs. Soon
neurons will web each body,
spreading fine mesh
through muscle and skin.

First, the neural folds
must fuse. Crest cells
edging a moon-bald field
reach with bulbous arms;
flowing inward, they move
toward each other.

And when they finally meet,
melding together, cell by cell,
there is no explanation:
they know who they are.
I can almost hear them
yammering in strange tongues.

Lucille Day

Tumor

Small and flat,
I inhabit the mind—
a landscape of pale stars
and electric trees,
where mountains fold on fields,
pearl on grey.

I crouch between cells
by the great salt lake
of the lateral ventricle.
All day I hear
the hiss of blood
twisting through tissue.

I begin to sing. I am
a tiny siren
calling the capillaries
to my cove. They come—
red serpents
coiling around me.

Oh how they love me!
Bringing their gifts
of food and nectar. I fatten,
slowly at first,
then faster and faster, until
I am round as a planet.

My cells spread everywhere,
each one a seed of another self,
but the skull closes
over a desert
tangled with old roots.
Sand starts to blow.

Lucille Day

Surgery

When a doctor leans over a body,
He must know what he means.
He must slip into a patient's lungs
Between the short, sick breaths.

He must pray he has washed enough,
That even his mind is empty of particles.

He must have a glass heart
That pumps only when he wants
And pulls him forward, fearless
Like a smart, expensive horse.

He must pray that he has slept enough,
That he is not hungry or thirsty.

His hands must move like milk
From a jug poured by a mother,
Motion more solid than bone.
They must fool even themselves in strength.

He must pray nothing jostles the earth,
Or take any unplanned breath.

He must not think of the damp sweat
That will take him later, in sleep,
But knowing where heat is lost,
He must keep his skin whole, cold,

And when the last cut tightens its lips,
He pulls the bloody second skin from his hands.
He sits touching the released flesh finger to finger
And holds back, holds back, like a closed church door.

Carol Burbank

The stethoscope

Through it,
over young women's abdomens tense,
I have heard the sound of creation
and, in a dead man's chest, the silence
    before creation began.

Should I
pray therefore? Hold this instrument in awe
and aloft a procession of banners?
Hang this thing in the interior
    of a cold, mushroom-dark church?

Should I
kneel before it, chant an apophthegm
from a small text? Mimic priest or rabbi,
the swaying noises of religious men?
    Never! Yet I could praise it.

I should
by doing so celebrate my own ears,
by praising them praise speech at midnight
when men become philosophers;
    laughter of the sane and insane;

night cries
of injured creatures, wide-eyed or blind;
moonlight sonatas on a needle;
lovers with doves in their throats; the wind
    travelling from where it began.

Dannie Abse

The Doctor Rebuilds a Hand

for Brad Crenshaw

His hand was a puppet, more wood than flesh.
He had brought the forest back with him: bark, pitch,
the dull leaves and thick hardwood that gave way
to bone and severed nerves throughout his fingers.
There was no pain. He suffered instead the terror
of a man lost in the woods, the dull ache of companions
as they give up the search, wait, and return home.
What creeps in the timber and low brush
crept between his fingers, following the blood spoor.
As I removed splinters from the torn skin
I discovered the landscape of bodies,
the forest's skin and flesh. I felt
the dark pressure of my own blood stiffen within me
and against the red pulp I worked into a hand
using my own as the model. If I could abandon the vanity
of healing, I would enter the forest of wounds myself,
and be delivered, unafraid, from whatever I touched.

Gary Young

77

The Parents of Psychotic Children

They renounce the very idea
of information, they are enamored
of the notion of the white tablet.
Their babies were outrageously beautiful
objects exploding their lives,
moving without compensation
because of them to worlds without them.
They believe they were presented
inadequate safeguards, faulty retribution,
and a concerted retirement into crime
of the many intent on their injury.
No two can agree on the miraculous
by which they were afflicted,
but with economy overcome
their fears of the worst. Their children, alas,
request nothing. And the far-fetched doctors,
out of touch with the serious truth,
are just practical and do not sing,
like the crazy birds, to their offspring.

Marvin Bell

Counting

The bulls and cows roam white
black, yellow, dappled
over the plains of Thrinakia
oblivious to the boys who fail
counting them all in one day.

The boys marvel at how the old ones
never seem old enough to die,
or how at dawn, light reveals
infinity over the herd,
as when wheat in a broad field moves
with one grace of shadow and light.

And entering the herd, the boys resume
what will never end, a problem
given by Archimedes centuries ago:
find the number of bulls and cows
of each of four colors
or find eight unknown quantities.

Each evening, the dust rising
from hooves, roseate with light,
hangs over the city. Already the boys
have returned home, while all around
the counted mix with the uncounted
multiplying tomorrow's unnumbered.

Michael Collier

Finnair Fragment

Ice berglets, poked down
by my oil rig stick
in Iitaala's fluted glass
fail to break the
roiled golden mirror
of jazzy bubbly, covering
a fleeting rift
of the laws of physics.
They really do like ice here . . .
Rise, perforated cubelets,
relent, let Archimedes
rest in peace.
Or have you, polyvowelled friends
conspired in brief white nights
to make a truly light champagne?

Roald Hoffmann

Telephone Ghosts

*"Large-scale radio communication on Earth
has been in operation for only some 40 years.
We may imagine those earliest radio trans-
missions — for example, a cadenza sung by
Enrico Caruso — travelling forever at the
speed of light across interstellar space from the
position the Earth was in some 40 years ago."*

Carl Sagan, Intelligent Life
in the Universe, *p. 393*

A number is dialed, a window opened;
the urgency races the currents like a silver pinball,
pulsing to some complex brain of copper nerves
and lead synapses.
The only thing that exists is
a small oval of intense
air.
There the ear is attuned to another universe,
an inverted world,
in a timeless moment of waiting;
the way a Klein bottle turns topographically
back into its own internal space.

Inside that static sea,
waves breathe;
waves hiss.
You hear a voice talking very far away;
and another behind that
like a diminished echo,
or that double mirror image,
shrinking to infinity,
which old barber shops had set up with their walls.

They are telephone ghosts,
phantasms that recycle and recycle
for centuries perhaps,
as electrons must in their atomic tracks,
whispering of mysteries in a different calculus.

Then there is a thorn-sharp click
and a greeting.
Someone speaks
and a part of her is lost,
severed from her mouth.
She speaks and one door
of communication
opens,
while one window,
paned in silences and whispers,
closes,
and she is reduced
by another fraction of her identity.
She is ghosted.

Addition, subtraction and division;
sound must bend to the mathematics of particle physics.
We rely more and more
on telephones
as our voices,
trapped into the lines,
try to sing back to us
with a tongue that is empty and heavy
as a cairn.

Robert Frazier

Thermometer Wine

Always hung on its plaque
on the porch like a mounted
icicle, but was so old
already the painted numbers
were peeling and hard to read.
Only Daddy could tell
the measurements — he'd known
the instrument since a boy.
At ten below it really
meant twenty, being slow
with age he said. At
ten above it was roughly
accurate, but on a hot day
he added twenty to its reading.
I watched the red needle
rise in the dog days and
marveled how the tiny
hair was both sensitive
and significant.
The blood rose in that stem
just a capillary of
bright, as though the day
were sipping through its ice
straw that special wine
and about to taste the
color from the drop at
the bottom that never clotted
or dulled no matter how
far up or down it wrote,
always chilled as snake or worm.

Robert Morgan

Entoptic Colours (1817)

for Julie

Let me tell you what with mirrors
Physicists of ours have wrought:
Some find phenomena delightful,
More torment themselves with thought.

Mirror facing facing mirror,
Doubles—exquisite effect;
Between them in the shadow stands a
Crystal cube which will reflect,

Analogue of earth, when mirrors
Glance their light across, the loveliest
Play of colours; and their twilight
Touches feeling, manifest.

Black you'll see, as black as crosses,
Peacock eyes you'll surely find,
Fades the light of day and nightfall,
Leaving not a trace behind.

And the word becomes a symbol:
Soaked with hues the crystal is;
Eye to eye we see like marvels—
That mirroring resembles this.

Let it be, the macrocosm,
With its phantom figurings!
Worlds we love and which are little
Harbour the most glorious things.

> *Johann.Wolfgang von Goethe*
> (translated by Christopher Middleton)

The Life of Particles

Each atom is an idea. Atoms are the floating part of all material, the part that makes you sit up suddenly and, smiling, say: "You know, suddenly, for some reason, I feel like going for a swim!" Actually the space in which atoms drift is dry, I only said that because they have this concept regarding floating. Actually, to speak objectively, what atoms are most pleased about are the little things, such as the fact that there is room in wood. That is the kind of space in which they dwell. Everywhere they look, there is a kind of atomic living-room. Atoms float around even inside the dust motes, that are like galaxies you see there floating around in the sky in the bathroom. Illuminated by a sun-ray that slips with the viewer's eye or the voyeur's eye through the tiniest crack in the venetian blind or louvre or whatever you use to cover the window, atoms have a life of their own. Every single atom is an idea, with a story to tell. Listen now to the light conversation in the dust particles. Is that just for the sake of politeness and sociability?—is it *really* what the people of the particles really feel? Here comes a recitation now, it is an atom in some poet's hand that is talking. Yes!—I, too, write my poems by hand and everybody reads them as if they were about what they seem to be about, only this is my only poem that is about what it seems to be about, this is my honest opinion, this is what it really feels like, this is my true poem at last, this is a message from an atom in my hand.

Michael Benedikt

Although in a Crystal

Although in a crystal
salt perhaps a topaz
my fixed point
> *(ignoring the fractures*
> *with their eternal light)*
becomes the ambience
and I float
disappear in the swells
of waves
or between particles of light
reappearing in
tubes of branches.

Or under a flat rock
slowly lifted
sucks air
from the bog.

Anselm Parlatore

Light in the Open Air

pieced from The Nature of Light and Colour in the Open Air, *by 19th Century Dutch physicist M. Minnaert*

PART ONE

Everything described in this book
is within your own powers of understanding and observation.

The Dark Inner Part

The Bright Outer Part

The Variability of the Colour of the Blue Sky

> Try to imagine that you are looking at a painting
> and admire the evenness and the delicacy
> of the transitions.

The Blue Sky

> What is the explanation of this curious phenomenon?

> What can be the cause of this wonderful blue?

> Compare the blue of the sky
> with the skies of Italy during your holiday.

The Colours of the Sun, Moon and Stars

> It is difficult to judge the colour of the sun,
> owing to its dazzling brightness. Personally,
> however, I should say it is decidedly yellow. . .

Be very cautious during these observations!
Do not overstrain your eyes!

Study the illumination. . .

Compare the light. . .

Compare the illumination. . .

Compare the luminosity. . .

Compare the light inside and outside a wood.

Intercept one of these images by a piece of paper.

PART TWO

Above all study your surroundings intently.

Move your opera-glasses gently a little to the left,
then to the right, and back again to the left. . .

> The Colour of Lakes

> The Colour of Puddles along the Road

> Strong Wind Rising, Grey Sky

One more reason to keep our eyes open.

Which are the feeblest stars
perceptible by you?

How dark is this shadow?

Practice on cold evenings. . .

Try to fill this gap.

Notice how much better you succeed with a little practice!

Estimate the strength of the ash-grey light
on a scale from 1 to 10. . .

Watch the separate breakers along the shore. . .

Imagine a pool of water in a hollow of the dunes. . .

>   Be sure to carry out this experiment;
>   it is as convincing as it is surprising!

Examine systematically the colours of the shadows!

Draw up a scale
for the phosphorescence of the sea!

Everything is meant to be seen by you and done by you!

Always try to find the explanation.

This causes a very peculiar sensation,
difficult to describe.

Annie Dillard

Wings

We have
a microscopic anatomy
of the whale
this
gives
Man
assurance

                    William Carlos
                         Williams

We have
a map of the universe
for microbes,
we have
a map of a microbe
for the universe.

We have
a Grand Master of chess
made of electronic circuits.

But above all
we have
the ability
to sort peas,
to cup water in our hands,
to seek
the right screw
under the sofa
for hours

This
gives us
wings.

          Miroslav Holub

Connoisseur of Chaos

I

A.   A violent order is disorder; and
B.   A great disorder is an order. These
Two things are one. (Pages of illustrations.)

II

If all the green of spring was blue, and it is;
If the flowers of South Africa were bright
On the tables of Connecticut, and they are;
If Englishmen lived without tea in Ceylon, and
        they do;
And if it all went on in an orderly way,
And it does; a law of inherent opposites,
Of essential unity, is as pleasant as port,
As pleasant as the brush-strokes of a bough,
An upper, particular bough in, say, Marchand.

III

After all the pretty contrast of life and death
Proves that these opposite things partake of one,
At least that was the theory, when bishops' books
Resolved the world. We cannot go back to that.
The squirming facts exceed the squamous mind,
If one may say so. And yet relation appears,
A small relation expanding like the shade
Of a cloud on sand, a shape on the side of a hill.

IV

A.   Well, an old order is a violent one.
This proves nothing. Just one more truth, one more
Element in the immense disorder of truths.
B.   It is April as I write. The wind
Is blowing after days of constant rain.
All this, of course, will come to summer soon.
But suppose the disorder of truths should ever come
To an order, most Plantagenet, most fixed . . .
A great disorder is an order. Now, A
And B are not like statuary, posed
For a vista in the Louvre. They are things chalked
On the sidewalk so that the pensive man may see.

V

The pensive man . . . He sees that eagle float
For which the intricate Alps are a single nest.

Wallace Stevens

Ways of Seeing

1.

Waves that were random
come on for a run,
organized Indians.

They press from a far place
helping each other's
thousand mile crawl.

One of them, your wave,
passes while you watch—
then just the ocean,

On and on and on.

2.

Leaves make a trail
lost people walk on,
and then a path no leaf
ever touches, where the hero
can wander guided by nothing,
a way of knowing where
the right path is
when a bird calls "Turn"
and the hero says, "No," and
the leaves move away
falling on all but the
unmarked path no one can find.

William Stafford

For Whitman

I have observed the learned astronomer
telling me the mythology of the sun.
He touches me with solar coronas.
His hands are comets with elliptical orbits, the
excuses for discovering planet X.

Lake water shimmering in sunset light,
and I think of the whitewashed dome of discovery
hovering over the landscape
wondering what knowledge does for us
in this old and beautiful un-knowing world.
                                        Yes,
                                            I would

rather name things
than live with wonder
or religion.

What the astronomer does not understand about poetry
is the truth of disguise.
That there are many names for the same phenomenon.
Love being
the unnamed/
the unnameable.

                                    Diane Wakoski

OBSERVATIONS OF THE
NATURAL WORLD

Getting Through

I want to apologize
for all the snow falling in
this poem so early in the season.
Falling on the calendar of bad news.
Already we have had snow lucid,
snow surprising, snow bees
and lambswool snow. Already
snows of exaltation have covered
some scars. Larks and the likes
of paisleys went up. But lately
the sky is letting down large-print flakes
of old age. Loving this poor place,
wanting to stay on, we have endured
an elegiac snow of whitest jade,
subdued biographical snows
and public storms, official and profuse.

Even if the world is ending
you can tell it's February
by the architecture of the pastures.
Snow falls on the pregnant mares,
is followed by a thaw, and then
refreezes so that everywhere
their hill upheaves into a glass mountain.
The horses skid, stiff-legged, correct
position, break through the crust
and stand around disconsolate
lipping wisps of hay.
Animals are said to be soulless,
unable to anticipate.

No mail today.
No newspapers. The phone's dead.
Bombs and grenades, the newly disappeared,
a kidnapped ear, go unrecorded
but the foals flutter inside
warm wet bags that carry them
eleven months in the dark.
It seems they lie transversely, thick
as logs. The outcome is well known.
If there's an April,
in the last frail snow of April
they will knock hard to be born.

Maxine Kumin

Waiting for *E. gularis*

*"An African heron was found on the north
east end of Nantucket Island. . ."*
                              news release
"The sighting of the century. . ."

                    —*Roger Tory Peterson*

Exile
by accident
he came

against
all instinct
to this watery place,

mistaking it
perhaps
as the explorers did

for some
new
Orient.

This morning,
dreaming
of the inexplicable

I rise from sleep, smoothing
the sheets behind me
to match

the water smoothed sand
silk
under my bare feet.

I walk past morning joggers
who worship in pain
the crucible of breath,

past dune and marsh
stockaded with eel grass
to this pond,

just as a breeze comes up
like rumours
of his appearance.

Young people in bathing suits
lounge here, fans
waiting for their rock star

E. gularis—even his name
becomes
an incantation.

The pond
is all surface
this cloudy day,

the dark side of the mirror
where nothing shows
until you stare enough

as at those childhood puzzles—
how many faces can you find
concealed here?

And there moving towards us
is the turtle's miniature
face,

and there the mask
the wild duck wears, stitching
a ruffle

at the pond's far edge
where now
the Little Blue herons

curve their necks
to question marks
(why not me?)

where in a semi-circle
ornithologists
wait

to add another notch
to their life
lists,

binoculars
raised
like pistols.

Linda Pastan

The Cormorant in Its Element

That bony potbellied arrow, wing-pumping along
implacably, with a ramrod's rigid adherence,
airborne, to the horizontal, discloses talents
one would never have guessed at. Plummeting

waterward, big black feet splayed for a landing
gear, slim head turning and turning, vermilion-
strapped, this way and that, with a lightning glance
over the shoulder, the cormorant astounding-

ly, in one sleek involuted arabesque, a vertical
turn on a dime, goes into that inimitable
vanishing-and-emerging-from-under-the-briny-

deep act which, unlike the works of Homo Houdini,
is performed for reasons having nothing at all
to do with ego, guilt, ambition, or even money.

Amy Clampitt

Camouflage

for Jo and Roy Shaw

It seemed at first like a piece of luck,
the discovery, there in the driveway,
of an odd sort of four-leaf clover—
no bankful of three-penny greenery
but a worried, hovering, wing-dragging
    killdeer's treasury—

a mosiac of four lopsided olives
or marbles you had to hunt
to find again every time, set into
the gravel as if by accident.
We'd have turned that bird's
    entire environment

upside down to have preserved them.
But what was there, after all,
we could have told her about foxes,
coons, cats, or the vandal
with its eye out for whatever anyone
    considers special?

In her bones, in her genes, in
the secret code of her behavior,
she already knew more than all our
bumbling daydreams, our palaver
about safeguards, could muster
    the wit to decipher:

how her whereabouts could vanish
into the gravel, how that brilliant
double-looped necklace could amputate
into invisibility the chevroned
cinnamon of her plumage. Cleverer
    than any mere learned,

merely devious equivocation,
that broken-wing pageant—
who taught her that? We have
no answer except accident,
the trillion-times-over-again
    repeated predicament

sifted with so spendthrift
a disregard for casualties
we can hardly bear to think of
a system so heartless, so shiftless
as being in charge here. It's
    too much like us—

except, after having looked so close
and so long at that casual handful
of dice, squiggle-spotted by luck
that made them half invisible,
watching too often the waltzing swoop
    of the bird's arrival

had meant a disruption of more usual
habits. For all our reading in the papers
about blunderers and risk-takers with
the shrug of nothing-much-matters-
how-those-things-turn-out, we'd unlearned
    to be good losers.

Sorrow, so far as we know, is not
part of a shorebird's equipment.
Nor is memory, of either survival
or losing, after the event.
Having squandered our attention, we
    were less prudent.

For a day, we couldn't quite afford
that morning's black discovery.
Grief is like money: there is only
so much of it we can give away.
And that much grief, for a day,
    bankrupted our economy.

               Amy Clampitt

Teaching about Arthropods

The male mite Adactylidium
hatches in his mother's body,
gobbles her up from the inside,
while mating
with all his seven
little sisters.

So, when he's born,
it's as if he were dead:
he's been through it all

and he's freelance,
in the bull's-eye,
in the focus
of extracurricular existence:

an absolute poet,
non-segmented,
non-antennated,
eightlegged.

Miroslav Holub

Brief Reflection on the Insect

The insect .is really not built too well. It needs
    a better skeleton, a better respiratory system
      and
    a better central nervous system than this
      couple of
    stupid knots. When improved along these
      lines,

    burying beetles can start public funeral
      services,
    scarabs rob banks, ants could be assigned
    to the space program and flies could supervise
    the whole world with big eyes, and make
      decisions,
    yes or no,
    good or bad,
    promote, punish.

The insect, safe from switching tails and DDT,
could reflect on the improvement of Man
who's really not built too well.

                        Miroslav Holub

In This Life

Now the cup of grasses and down is cool,
the eggs cool, the throne empty
from which she would step down and growing still,
look out long toward the darkness.

One moist and terrible night, it came creeping
and tensing its jaws and it too grew still
and then it had her,
and a small diamond of light opened in her brain.

I remember the eyes closed tight
against the final ecstasy of the teeth,
the weightless blood-beaded lump of feathers buried now
under the iris slowly eaten alive by the air.

Here is the father
blossoming on a twig
to sing the song of the bleeding throat
on this day of crystal wind and young sunlight.

He sings the endless song
of irises wrinkling and wrinkling and becoming nothing,
road of fine sand strewn with fallen wings,
the mouse, the toad, the blind nestling taken in deep grass,

he sings of a diamond, sings
of the spokes flashing brilliance at the center
of the ceaselessly collapsing floor of bone
and I wish I sang with him,

he sings the only truth
in this world where men remember mostly lies,
sings it and sings it
till it breaks at last into particles of light,

blossoms of mercy
in the midst of the holocaust.

Robert Mezey

Accommodation

In the deep sphagnum moss
of the bog
a marsh treader
fuselage hurrying
out onto the smooth water

six jointed legs dimple
the strong silk.

Under
an insect, this
a water boatman
leaves the stiff stem of a chara
sculls upward

music

before the treader spears it
sucks the body dry.

Anselm Parlatore

The Spider

His science has progressed past stone,
His strange and dark geometries,
Impossible to flesh and bone,
Revive upon the passing breeze
The house the blundering foot destroys.
Indifferent to what is lost
He trusts the wind and yet employs
The jeweled stability of frost.
Foundations buried underfoot
Are forfeit to the mole and worm
But spiders know it and will put
Their trust in airy dreams more firm
Than any rock and raise from dew
Frail stairs the careless wind blows through.

Loren Eiseley

Winter Sign

A spider web pulled tight between two stones
With nothing left but autumn leaves to catch
Is maybe a winter sign, or the thin blue bones
Of a hare picked clean by ants. A man can attach

Meanings enough to the wind when his luck is out,
But, having stumbled into this season of grief,
I mean to reflect on the life that is here and about
In the fall of the leaves—not on the dying leaf.

Something more tough, reliable, and stark
Carries the blood of life toward a farther spring—
Something that lies concealed in the soundless dark
Of burr and pod, in the seeds that hook and sting.

I have learned from these that love which endures the night
May smolder in outward death while the colors blaze,
But trust my love—it is small, burr-coated, and tight.
It will stick to the bone. It will last through the autumn days.

Loren Eiseley

The Beekeeper's Dream

They go before the fog lifts, looking for light.
The workers dance to tell the hive the way.
The sun directs the angle of the flight

the dance describes: they circle, waggle, kite
out in ecstatic arcs. Keen on the honey's fine bouquet
they go before the fog lifts. Looking for light

in Brooklyn's ailanthus trees, these cosmopolite
creatures seed the city green on their foray.
The sun directs the angle of their flight

or else the memory of the sun on a familiar site
—a tree, the Gardens, or the Bay—if it's a cloudy day.
They go before the clouds lift, looking for light.

Tired in the evening, they return legs loaded, bright
with pollen: daffodil yellow, rose red, grey.
The setting sun directs the angle of their flight.

Colliding with hoodlums and a knife last night,
I was stripped, stung, robbed for pleasure. I pray
they'll go before the fog lifts, looking for light.
The sun directs the angle of their flight.

Katharine Auchincloss Lorr

Amphibian

End over end, a leaping
anecdote: aloft among
the plucked trajectories
of insects, it becomes
    a plummet

down the bole-walled
sunlight's cistern,
to lodge, an avatar
of the New Hampshire
    demiurge,

a lapidary throwback
to ancestral ponds'
stunned immobility:
two seemingly unsee-
    ing slits

of agate set in seem-
ing stone, along the
flank a gilded glyph,
beneath the haunch
    the flutter

of what otherwise
appears to petrify
declaring everything
to be intractably
    in motion:

as we look, a leap, a
quivering in absentia
where motion's once
immobile artifact
    had been.

        Amy Clampitt

Pit Viper

A slow burn
in cold blood
is all snake
muscle does.
The nerves drone
their dull red
test pattern
for days. Days.

The eyes, black
pushbuttons,
are just that.
On each side
a fixed dish
antenna
covers the
infrared.

The ribs, like
a good set
of stiff twin
calipers,
lie easy
and don't take
measure of
what's not there.

The skin's dim
computer-
controlboard
arrangement
of massed lights
displays no
waste motion.
Take no joy.

A strike force
is no more
than its parts
but these parts
work. Dead-game
defensework
specialists,
they die well.

The point is
they don't choose
livings: they
don't choose Death.
God save them
they aren't small-
time haters
that joined up.

George Starbuck

At the Scenic Drive-in

There weren't any other cars —
it was late & cool — & the cook himself
brought our orders out.

Behind us mountains seemed to whisper
as the sun balled down
to flare wide on the ocean;

when he returned to collect our tray
we murmured politely
about the beauty of his spot.

"That superhighway takes the traffic,"
he said, "so now I can't afford
either lights or gas,

but if you like to look, go ahead,
stay on long as you want
& count the stars too."

Immediately I wished
an earthquake would swallow him,
my temperament is so bloody;

but I just said thanks, & we kept on
watching red-refracted solar light
deepen to a dark filter

through which clusters, constellations,
galaxies — if nothing else —
somehow pass.

David McAleavey

Equinox

Today the light will be as the darkness; our dreams
and what we have learned to expect in their place
will be equal. The sky senses this and divides
into enormous strips of white cloud and blue.
A breeze is in from the ocean, folding the branches
and leaves of the black walnut toward the mountains.
Monarchs hang like windows of stained glass from the fruit trees.
The world cannot be more perfect than it is today,
when what we have gained is equal to all we will lose.
The sun is setting behind the eucalyptus and already
a harvest moon rises. Tomorrow will be a shorter day.
Below me, around the table my feet dangle beneath,
alyssum and wild mint strain toward the weakening light.

Gary Young

Once Only

almost at the equator
almost at the equinox
exactly at midnight
from a ship
the full

moon

in the center of the sky.

Sappa Creek near Singapore
March 1958

Gary Snyder

At Camino

Tonight, early August
and a river of stars is falling
down the long arms of the Milky Way.
In the northeast
and faintly visible if you're lucky,
the Andromeda Galaxy rises in its slow circle
toward the equator
and on to its setting, months from now,
following Cygnus down to the west.
We have been granted another year,
the first breath taken and gone
while above, a meteor, luminous and brief,
is accepting and releasing our piece of light.

                                    Timothy Sheehan

Tornado Watch, Bloomington, Indiana

The sky, reading our thoughts
as always, blackens and grows still.
I am filled with the menace
of my own mind, of bad weather
when my chest spasms and I think
gunshot as the lightning
blinds me, and the room shatters
with the nearness of it.
The clock, the kitchen table,
the dishes swimming in their rack
pulse and glow; a purple light escapes
everything and the moment will not pass.
I cannot breathe but the air
is sweet, thick, and seems to
push its way into me.
If I could move, I would not.
Dying, suspended within us and held close,
could be like this. In my terror
or pleasure, I imagine heaven, until
from outside, a starling in the sycamore
begins his hoarse song and I go on
living in the durable world.

Gary Young

At Liberty

Snow lies in claws
on each plank of the porch.
Thin and fierce at the tips.

Preening, prodigal
in their brillance,
these crystals flex,
bend the light on the wind.
Prisms of a self
complete as a fly's eye,
they live off light,
give no heat.
Their harshness fixes us.

As our bodies curve in a frozen arc
toward them
(only our breath is warm)
their dazzle darkens.
They melt.

Anne S. Perlman

From a Rise of Land to the Sea

The water's shore-lapping signature
is a random drone, picking a
wet string of nature's scrapping still
moments. Sun-freckled wavelets dive.
Yawl rubs against buoy, teased
to a sporadic dulled tinkle that
rises over the wind in the
lindens. The same wily actor
folds the feel of the sea gently
into my back, drives the clouds.

The multisensual mixing is darned good, my engineer, my director.
You even provide low comedy in a pesky fly and drama
in the jet swish of a swallow diving to her
eaves nest that I, intruder here, obstruct.

Roald Hoffmann

Crossing

It was evening when we came to the river
with a low moon over the desert
that we had lost in the mountains, forgotten,
what with the cold and the sweating
and the ranges barring the sky.
And when we found it again,
in the dry hills down by the river,
half withered, we had
the hot winds against us.

There were two palms by the landing;
the yuccas were flowering; there was
a light on the far shore, and tamarisks.
We waited a long time, in silence.
Then we heard the oars creaking
and afterwards, I remember,
the boatman called to us.
We did not look back at the mountains.

J. Robert Oppenheimer

Rivulose

You think the ridge hills flowing, breaking
with ups and downs will, though,
building constancy into the black foreground

for each sunset, hold on to you, if dreams
wander, give reality recurrence enough to keep
an image clear, but then you realize, time

going on, that time's residual like the last
ice age's cool still in the rocks, averaged
maybe with the cool of the age before, that

not only are you not being held onto but where
else could time do so well without you,
what is your time where so much time is saved?

A.R. Ammons

Brain Coral

```
Waves                                    African wind
      of                         the hot
         Sahara sand          upon by
                  breathed
are long                              rise and
        black waves                 the
               of a       hair are
                  lover's
fall of                                    graph paper
      hips are                 and white checked
            a sine     upon blue
               wave
are                      magnet.
   iron                of the
        filings   adoration
               in
```

Harden these waves like a heart into a mind
and I hold in my hand a hemisphere
of limestone silence.

Lois Bassen

Whatever It Was I Was Saving for My Old Age

What was it I was saving for my old age?
Whatever it was, I'd better get to it now.
Was it collecting? or mystery novels? or
gardening? I've gardened already, rows of
peas, beans and tomatoes surrounded by
nasturtiums to keep away the cut-worms,
dazzling nasturtiums in all colors, far
better than any I ever grew trying to
grow flowers for their own sake. There
is a lesson here somewhere: the tomatoes
kept the nasturtiums free from aphids? was it?
tiny little black bugs that look like grains
of pepper moving, or under-grown caviar pasted
along a stem, succulent and bending. And
the nasturtiums with their own peculiar
pungency put up an invisible shield
to turn the worms from the tomatoes.
The point being that two different
plants, one flower, one fruit (both
edible in part — nasturtium leaves
are tasty salad greens) with a built-in
unique aroma (for that is what the shielding
seems to consist of — strong smell
which each small nose or whatever they use
to smell with cannot cope with) still,
these two plants, unrelated in any
other way, could, standing side by side,
protect the other from its natural
enemy, the unrooted creature which could
eat them up, worm, bug, whatever.

I am trying to make the leap from plant
to human but I seem to be swaying on
the stalk of a nasturtium flower much
too fragile to hold my weight. I was
trying to make an analogy: about homo
sapiens and love, a comment on symbiosis,
but I keep swaying here on this nasturtium
stem and what I seem to have saved for my
old age is science fiction.

Ann Darr

Boundaries

Floating finned and masked,
Sinking blueward
Into these wilds
Where icons dispose of dry senses:
First, the air goes, then gravity,
    And finally time.

Diving down the banked corals
I brush a polyp cluster with one bare arm,
Bruising its delicate loops and rows.
This coral boulder carries life at the edge
Blooming on the lately dead.
So, here is another animal
Making bones from its own soft body.
Here is another animal
With surprisingly thin skin.
Using the same solutions, our various
Bodies seal the ocean from the flesh:
Our skins hold the line against the sea.

Others have different styles:
Boning up their surfaces,
As obviously as snails.
Still using Cambrian ingenuity
To suck calcium from the sea
and spin it out as shell—that trick
On which bone making trades.

We carry our boundaries
In ancient rank,
Inheriting not only ocean
Blood, but also bone.

                Carrol B. Fleming

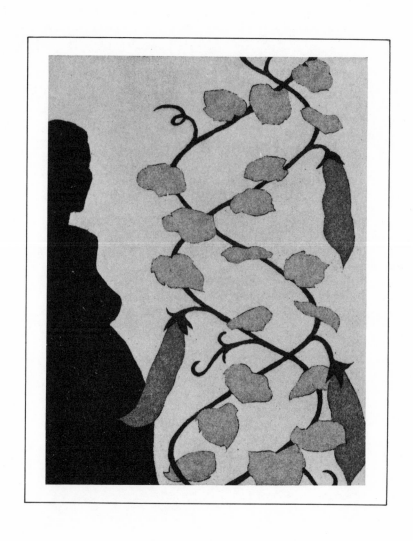

SCIENCE AS METAPHOR

Mendel's Law

I

A monk can do his work on bended knees
inside or out; the bishop looked askance
when Mendel labored in a row of peas
and led the combinations in their dance.
The spark of genius dominates the heavens
and sparkles in the furrow and the loam;
both earth and sky are broken down in sevens
and Christ is captured in a chromosome.

My lover, this was many years ago.
Mendel became abbot and then died.
But all his scorned experiments proved so:
the row of peas spoke truth, the bishop lied.
And what has this to do with us? I'm wild
to know it all since you are now with child.

II

The double helix and the triple star
work in conjunction, like harmonic tones,
and I will praise—how beautiful you are!—
the spiral staircase turning through your bones.
Genetic links, for better and for worse,
bind us to all creation: in my ears
your voice has blended with the universe
and vibrates with the music of the spheres.

Your fingers on the keys at Christmastide,
so effortless in their precise selection,
pick out the ancient chords; while I, beside
you, turn the pages at each soft direction
and wonder at your slender hands because
your fingers follow God's and Mendel's laws.

III

When Eve was cloned from Adam's rib, and stood
by the serpent underneath the Tree
she understood what lovers understood
since first they separated from the sea.
Her choice was meagre; still, she had to choose;
and we, like Eve, have chosen ever since,
face to face, the brown eyes to the blues:
it is the choosing makes the difference.

And in the code that Mendel labored on
our child will be deciphered; there will merge,
in childish shape and spirit, a paragon
where paradox and paradigm converge.
Now I can see Eve's children in your eyes:
completely new, yet linked to paradise.

Peter Meinke

Little Cosmic Dust Poem

Out of the debris of dying stars,
this rain of particles
that waters the waste with brightness;

the sea-wave of atoms hurrying home,
collapse of the giant,
unstable guest who cannot stay;

the sun's heart reddens and expands,
his mighty aspiration is lasting,
as the shell of his substance
one day will be white with frost.

In the radiant field of Orion
great hordes of stars are forming,
just as we see every night,
fiery and faithful to the end.

Out of the cold and fleeing dust
that is never and always,
the silence and waste to come —

this arm, this hand,
my voice, your face, this love.

John Haines

Self-Portrait with Hand Microscope

After the Expressionists

The room is red
for the rooms of my heart
and the births of children.
I am the thin lady
(stiff, angular brushstrokes)
in denim and silk.

Half my mouth curls
in a cynical grin;
one eye is weeping.
I hold the microscope
between thumb and forefinger
like a silver pen.

Having looked at onion roots
and rats' eyes, I bend
to examine my skin,
its star-burst pattern
cracking slowly,
my man in shadows

where mildew and bread mold
are beaded, intricate.
Earthworms cling
to his heart. O slick
muscle closed like a fist!
I search my mind

for the beat of cilia,
their synchronized dance,
the sudden wings
of a fruit fly,
flickering—movement,
live things.

                    Lucille Day

Nystagmus

1. Drowsiness, 2. A rapid and
involuntary oscillation of the eyeball.

Slow to wake, often I catch
just a fragment of you,
a bright trail of color
on your way to work. Only
the clock fights my drowsiness,
the second hand's quick circles
holding the eye. The rest
is too static; the familiar
dissolves to a homogenous whole,
books, wall, rug, door.

Like hummingbirds, the eyes
vibrate continuously, darting,
blurring vision slightly. If an image
is projected to it moves with the eye
it will be seen, for a moment, completely stable,
clearer than ever. Then the edges
begin to unravel, and the image
fades and disappears. This vibration
makes the world move,
keeps it there.

Some days I look up and see you
casual and perfect as Giotto's 0,
and a panic overwhelms me,
though you are almost
speed itself, and my hands grow icy,
want to grab you and push,
keep moving! keep moving! I stare
too long at the things I love, could not
bear my eyes to break you apart.

Joseph Matuzak

from Backgammon

Long history
Volcanic activity in the solar system
Shiny green outfit of some kind
Rare supernumerary arcs
You can have what you want
A little room called love's valley
700 × 200 feet glowing red at night
Fast wink
Slow promise
The dance is the whole big mouth
Rocks on the desk
Shells windowsill
Each scratch on the brain
Ouzo
Honey
Greek coffee
Worry beads
Oregano
Sweetie
Rugs
A snatch and snatch: joyful moments
Light lapping pebbles
Flat narrow boats
Red yellow stripes
The dazzle of horizon
Poetry is not good
But the desire to poetry
The blue equivalent hills
Of course I love vibrato
Although to dance with god
Is to dance with ourselves
Said the physicist
Violent love so slow it cuts
Cage from around the heart

Olga Broumas

Perpetual Motion

They're changing partners again, safely unseen
(Or so they thought) on the other side of the wall
Where death, dialing the defunct
Phone numbers you still know by heart,
Reaches an eternal dial tone.

Time out for a few last questions.
"Is there a unifying principle to these kisses and betrayals,
Some heavenly conspiracy that controls such accidents
So that they seem to make sense? Or are the players
Just pieces in a Jackson Pollock jigsaw puzzle?

"Does the malevolence of nature console us
Because, though innocent, we have never been good,
Or do we recoil in horror
From the grinning clown face on the back
Of a cobra's extended hood?

"No wonder we feel misunderstood.
You can measure our velocity but not our location
As we round the curve into the recent future,
Afraid to say what we have seen,
Alone and together on the way."

David Lehman

A Physics

When you get down to it, earth
has our own great ranges
of feeling—Rocky, Smoky,
Blue—and a heart
that can melt stones.

Miles above, some still pools
fill with sky, as if aloof.
And we have eyes
for this—the earth and its
reminding moon. We too

are ruled by such attractions—
spun and swaddled, rocked
and lent a light. We run
our clocks on wheels, our trains
on time; but all the while we want

to love each other endlessly—
not only for a hundred years,
not only six feet
up and down. We want the suns and moons
of silver in ourselves, not only

counted coins in a cup. The whole idea
of love was not to fall. And neither was
the whole idea of god. We put him well
above ourselves because we meant,
in time, to measure up.

Heather McHugh

Waiting

for my father

His thesis was crystals
compact enough to live
for weeks on sunlight.
He showed us constellations
while we slept on pillows
under meteor showers,
waiting for supernovas.
Or did headstands
on wet grass under the moon's
partial eclipse, pink craters
that ached to be beautiful.

Once in Virginia the sun
performed for him,
with shadow waves,
the threat of total blindness,
the diamond ring effect.
Always his telescope,
the extension of brown myopia,
was poised under the stars
he owned, the red and the blue,
the Horsehead Nebula,
the Seven Sisters.

We are still waiting
under starless nights for bright
and focused points of light,
with names like Ganymede, Io,
and Cassiopeia,
to count more moons,
to fall asleep under the influence
of a different gravity.

                    Judith Skillman

The Causes of Color

(from an article in Scientific American*)*

1

Natural alexandrites are very rare,
red in candlelight
blue-green in the sun.

The color is generated by
chromium ions — impurities.

Impurity has always given me
my color: big eyes, long nose,
reading too many books.

2

Chromium ions are sensitive to their
chemical environment.
They make the ruby red,
the emerald green.

The ruby and emerald are similar.
You say red, I say green.

3

An unattached electron
gives rise to color.
So can the absence of one of a pair
— a 'color center.'

Better to see oneself as
an unattached body
than half of a broken pair.

4

The glass of some old bottles
contained iron and manganese.
Years of sunlight caused color centers,
the bottles became desert-amethyst glass.

But excessive heat
can destroy the color center,
and the color.
A hot amethyst turns citrine quartz,
or the rare greened quartz.

The heat of contained anger
can destroy a woman's center.
She appears ordinary
until it is released.

5

In semiconductors
the band of energy levels
is split in two.
Between the band of lower energy levels,
and the band of the more excited
is the gap of forbidden energies.

My anger is forbidden energy.
The largest gap between
what I feel and what I say
renders me colorless.

6

When wavelengths interfere
some amplify each other, others
cancel each other out.
The swirl of oil on water,
the gleaming beetle.

Our pasts interfere.
When hurt, we amplify our pain
and cancel each other out.
Yet we stay for the warmth,
the background gleam of joy.

7

In diffractive grating,
light passes through
equally spaced points or lines.
The light is scattered,
and waves from the openings interfere.
My mother's opal is a grating.
She loved it before it was understood.

Ann Rae Jonas

Migration as a Passage in Time

This afternoon,
unsponsored and diminishing, is
like our futures: limited
but not to be dismissed.
Birds apron the snow.

Even now,
sea turtles are slipping
through the South Atlantic,
charted for an island drowned
a hundred million years.

The shifting continents
shot the Himalayas up
like so much seed,
gave air to archipelagoes
but cannot shape
the streaky thoughts of sea turtles
who swim a thousand miles for a nest.

Kite me across clouds, the oceans,
winter hills gone vacant
to a waxy sun—

our anticipation is unfounded
and as dreamlike
as a meeting on this winter night,
the mountain's face
spilling all its starlight into yours.

Jody Bolz

The Monkish Mind of the Speculative Physicist

A wisp of slight sound, an echo of an echo
through the trees—or something imaginary
     projected by

his fevered brain. Bilaniuk examined long
into the night the properties of
     imaginary mass:

according to Relativity Theory he found
that at one-and-a-half times the speed
     of light a rock

which weighed in this world a pound would weigh
the square-root-of-minus-one pounds.
     The number is called

Imaginary, not impossible, and if, he said,
a particle could go that fast it would only
     slow down

the more it were pushed. An Adam who conjured
new creatures to name, he called them
     Tachyons. A slight

fever was the cause, and insomnia, nothing
really, just the influenza then in fashion.
     A dripping faucet,

a lack of stillness in the night becomes
a dread and catches on the tick of the clock
     as annoyingly

as silk on a hangnail. Meanwhile a woman
emerged into morning from a night
     of bad dreams.

She wandered into the small garden
and tried not to think of the children
    still sleeping

or the husband oblivious in bed. She watched
sunlight slice itself in the dew of an iris,
    considered how

it might be in some fair foreign country to live
with a dark stranger. She finds frightful
    her mirroring mind

    reflecting what cannot be

onto what cannot stay nameless forever.

Bin Ramke

Rorschach

In this ink blot, there are two lions
pulling meat apart, or two moths
fluttering against a streetlamp.
This one is a hieroglyph from a lost language,
it means *autumn* or *momentary*.
This is the veiled face of an Arab woman
afraid the camera will steal her soul.
And in this one, two people can't agree
how the melody went in a tune
they heard last night, or if they are in love.
And every time she says *I love you*
she is lying, but it is also a lie
whenever she won't say *love*.
In this ink blot, shadowy hands
have just let go a milkweed pod.
Seeds are falling everywhere.

Laura Fargas

Wavelength

They were sitting on the thin mattress
He'd once rolled & carried up the four floors
To his room only to find it covered nearly all
Of the bare wood
Leaving just a small path alongside the wall

& between them was the sack
Of oranges & pears she'd brought its neck
Turned back to expose the colors of the fruit
& as she opened a bottle of wine
He reached over to a tall stack of books
& pulled out *The Tao* & with a silly flourish
Handed it across the bed to her   she looked up
& simply poured the two squat water glasses
Half-full with wine & then she
Took the book   reading silently   not aloud
As he'd assumed & suddenly he felt clearly
She knew the way
Two people must come upon such an understanding
Together of course but separately
As the moon & the wave remain individually one

David St. John

Two Sorrows

He had lived for the sorrow of numbers
& this had made his mind beautiful
& also pure
          somewhat
Like a globe of red ink held up
In a beaker before the light of the setting
Sun by a woman in a white smock who
Without question desires him
If there is any

Equation he cannot yet complete

It may be that of red ink $\neq$ blood
Though it may also concern the ellipsis of
Sweat along her lips

Beading a bit like the light in the beaker

As he puts his hand around hers for only
A moment & the liquid swirls a little
In the bottom of its glass bulb
& he awakens quite suddenly beyond his dream
Of riverbeds erased by snow
An ostrich at her egg
A boy asleep in the high heavenly forest

Of innumerable & open arms

                              David St. John

Love's Alchemy

Some that have deeper digged love's mine than I,
Say where his centric happiness doth lie;
    I've loved, and got, and told,
But should I love, get, tell, till I were old,
I should not find that hidden mystery;
    O, 'tis imposture all:
And as no chemic yet th' elixir got,
    But glorifies his pregnant pot,
    If by the way to him befall
Some odoriferous thing, or médicinal;
    So lovers dream a rich and long delight,
    But get a winter-seeming summer's night.

Our ease, our thrift, our honor, and our day,
Shall we for this vain bubble's shadow pay?
    Ends love in this, that my man
Can be as happy'as I can if he can
Endure the short scorn of a bridegroom's play?
    That loving wretch that swears,
'Tis not the bodies marry, but the minds,
    Which he in her angelic finds,
    Would swear as justly that he hears,
In that day's rude hoarse minstrelsy, the spheres.
    Hope not for mind in women; at their best
    Sweetness and wit they're but mummy, possessed.

John Donne

Eclipse

Late May
and twenty miles east of Fresno
it is 95 and rising.
Last night
I watched the moon eclipse
going redder than grapes in the San Joaquin
until it disappeared,
then walked down a road between fields
where vines hung to the ground
brushing the tall grass.
When the moon went out
the planets hidden in the grapes
rose to my eyes
and perhaps it was you I saw then
walking into the field
and the abrupt dark.
Today it is the heat
and a woman who lets out her hair
as she walks into a river that slumbers,
and the sun keeps rising through our lives,
and the invisible stars.

Timothy Sheehan

Down, Down, Down

After seven months in space, the astronauts
touched down. We saw them trundled
from their second home—they couldn't hold
a bunch of flowers up. The world was heavy;

heavier, they couldn't bear
themselves. The story they had starred in
turned from Music of the Spheres
to Newsreels of their New Wheelchairs. So much

for weightlessness as grace. For months
after their landing, none of them
could sleep. And this is how
they put it: down is hard.

As for myself, I always had
a comforter or dream to sleep with, stars
connected by the best of lines. My life
was always looking up; I always had

a little always in my bag, and took
all loss for temporary. When
was my comedown? Did someone
disappear? Did I get old? There is no cushioning

the outcome; suddenly the future leaves
you cold. Sometimes I think that I can think
the worst away, telling myself that time
is weather in Marseilles; but time's not just

the element I'm thinking of, it's what
I'm thinking in, like having only mind
to grasp mind with. By then I've used up
maybe thirty of the eighty-seven thousand

seconds in a human day. It's sad.
However wisely I may weigh the evidence, or reconcile
the opposites (before and after both
are then) the world

is not so manageable. It will not
be held or had. And when the man I loved was gone
no hill or hole, no president or anchorman
had pull enough for that. My moons and oceans

don't mean anything to him. He won't come back.

Heather McHugh

A Curfew: December 13, 1981

Fever, the clang in the beleaguered pumproom
muffled with Tylenol, banked under icepacks,
rising while outside snow fell, a seeming flux
of strict constructions, the vapors' mimic
of turmoil among the leucocytes—mass panic,
blocked corridors, the riddlings of dispersal:
Why? If meaning is a part of any system,

what laws apply? To Alfred Wallace, burning
on his bed in the Moluccas, the malarial shimmer,
parting, whispered "Malthus": accident, disease,
war, famine certified as a severe epiphany, a random
elegance unfeeling as the Snow Queen's hex, the filter
of the future of the species. The stoic laughter
of Democritus: Nothing truly is except the atom,

the Whole a sieve of particles, its terrors
loomed of shadows' cumber. Along the thoroughfares
of Warsaw martial law, the day my brother died,
serried the pallid Baltic sun with roadblocks;
a curfew overtook the solstice. This winter, would
the sun turn round again for the gregarious gamble
of Solidarność? My brother dead, I cried over the news.

He'd looked into the murk of so much turmoil,
flux and rigor, unbought *piètàs* of the suicidal,
such jigsaw-fault-line fracturings of seeming
entity, fears of the action of God knew what laws
laid down by God knows who among the shadows
of the cave, the cloakroom or the bedroom—listening
head down, eyes impassive, musing, feeling his way

along the pillared halls of withheld judgment—
and still, like a despairing small-hour phone call,
they trailed him down the fever's passageways
into the pumproom of delirium. "I think I won't
go to the office for a while," he murmured. "From
now on, just a few private patients." The snow fell,
the fever guttered, and the streets of Warsaw froze.

The thinnest of osmotic boundaries contain what once
was called the soul; the universal laws, the flux
of Heraclitus, packed yin-and-yangwise into the globule
of the infinitesimal, are now coöpted for a game of jacks,
taws toyed with by the hubris of a species whose petulant
chevaux-de-frise infest the globe with roadblocks,
a raging mimic of the universe's grand indifference.

Amy Clampitt

80-Proof

A fifth of me's me:
the rest's chaser:
35 lbs.'s
my true self: but
chuck 10 lbs. or so for bones,
what's left's
steaks & chops &
chicken fat,
two-over-easy & cream-on-the-side:
strip off a sheath of hide,
strip out nerves & veins
& permeable membranes,
what's left's a greasy spot:
the question's
whether
to retain
the shallow stain
or go 100% spiritual
and fifth by fifth
achieve a whole,
highly transcendental.

A.R. Ammons

Hieroglyphic

Once a girl
innocently typing
put the wrong word
on the page
and in a flash
changed my career
from physiology
to psychology.

I thought: What if
the same girl types
the place of my birth
after I die—Baltic
for Baltimore
or writes Black Forest
for White Mountain
as my family name
or instead of Repose
puts down Poseur?

I thought: What if
the cause of death
were a broken heart
instead of a broken-
down heart? My last street
on earth Wormwood instead
of Marywood. Who would
argue my case for me?

Surely some good
professor would come
and read down the tome
of my life from top
to bottom like Hatshepsut's
hieroglyphs, making up
a wonderful story from wavy
lines and birds and small
unexpected footprints
erased by the sea.

Myra Sklarew

Song

All phantoms of the day
Nothing burns or pierces this dense fog
Ghost of the table
Ghost of the typewriter
Past and future mutilate the appetite

I have my name
For the whole shambling charade
Enter and exit
Tentative changeling imperious swineherd morning

Nothing imagines
A man a wife and children
Pieces of pure substance keep skittering away
Time sits high
On the throne of calcium

Debris of the void
Perpetual debut of the blind dancer
We dance blindly
Deafly
Even as the door opens
A little wind
Even as the oak
All the arms lift cups of light to the zenith
And the fiery brown earth
Speaks in tongues

Robert Mezey

Poem Technology

It is
    a fuse,
which you set off
somewhere in the grass
or in a cave,
or in a third-rate
        saloon.

The flame darts
past stalks
and bewildered butterflies,
past startled stones
and drowsy mugs,
darts,

spreads a bit
        or shrinks
as pain in a surplus finger,
hisses, sizzles,
stops
    in a microscopic vertigo,

but at last,
at the very end,
    it blasts,
a bang from a cannon,

crumbs of words fly
        through the universe,
the walls of the day rumble

and although
the rock's not cracked,
at least somebody says—

    Shit, something happened.

        Miroslav Holub

Natural History

for Adrienne Fargas

From earliest schooldays
children are shown
how small the earth is,
how short man's life
has been. If the sun
is a basketball, the earth
is a cherrypit in the next county.
If the planet is an hour old,
man has only been here
the wink of an eye. And that eye
would not see you, seven years old,
standing rapt in the museum
awed by dinosaur bones. It notices
only seas filling in, mountains
popping up. If the sun
is an orange, this poem
is bigger than the whole world.

But the world is bigger than you are,
wrapped warm and lonely
in the dark bed of night,
and the moon is invisible.
The moon is not even
an appleseed two doors down
yet the earth can never
coax it closer. Your own palm
blocked it out last night, then
waved reassurance: we won't
leave you out there alone,
you stubborn old white moon,
promise. It tugged at you,
dry and alluring, and dinosaur
bones today reawaken that
lift in your blood. But something
holds you down: you stuck out your tongue
at the dinosaur, closed your eyes
for longer than a wink
without disappearing, and the sun
stayed a basketball, bigger than
this poem, smaller than the whole world.

Laura Fargas

SATIRE AND CRITICISM
OF SCIENCE

Newborn Baby

With eyes like embers of an extraterrestrial
   civilization,
it occurs. Garbage in the wind.

    It asks:
    What about neurosecretion? Solved?
    Or this red shift of galaxies? Is it explained
       yet?
    Have we got malignity under control?
    Or at least the theory of aspirin effects?
    And the particle wave problem?
    Laws of thermodynamics, number four?
    And what about this crappy mess here?

The newborn baby, plainly disappointed, becomes
   engrossed in itself.
Gradually, it's covered by fine hair, and at night,
   almost imperceptibly,
it whines.

But the pack moves away.

Miroslav Holub

Doorman

The night we heard the news from space,
my daughter, who is three, remarks
with no surprise but careful to instruct:
"The moon is like a doorknob,"
to that other self all children seem
to have and have to answer to.

I sit trying to construct a poem of praise.
Spacemen and women stumble down the page.
She says, again, impatient to be gone,
"The moon's a doorknob," and,
already dressed to play outside,
waits for me to open up the sky.

Martin Galvin

La Guerre

O sweet spontaneous
earth how often have
the
doting

      fingers of
prurient philosophers pinched
and
poked

thee
, has the naughty thumb
of science prodded
thy

     beauty   . how
often have religions taken
thee upon their scraggy knees
squeezing and

buffeting thee that thou mightest conceive
gods
      (but
true

to the incomparable
couch of death thy
rhythmic
lover

    thou answerest

them only with

            spring)

                 e.e. cummings

When I Heard the Learn'd Astronomer

When I heard the learn'd astronomer,
When the proofs, the figures, were ranged in columns before me,
When I was shown the charts and diagrams, to add, divide, and
    measure them,
When I sitting heard the astronomer where he lectured with much
  applause in the lecture-room,
How soon unaccountable I became tired and sick,
Till rising and gliding out I wander'd off by myself,
In the mystical moist night-air, and from time to time,
Look'd up in perfect silence at the stars.

Walt Whitman

ARCTURUS is his other name,—
I'd rather call him star!
It's so unkind of science
To go and interfere!

I pull a flower from the woods,—
A monster with a glass
Computes the stamens in a breath,
And has her in a class.

Whereas I took the butterfly
Aforetime in my hat,
He sits erect in cabinets,
The clover-bells forgot.

What once was heaven, is zenith now.
Where I proposed to go

    Without a tighter breathing,
    And zero at the bone.

                  Emily Dickinson

True Enough: *To the Physicist* (1820)

"Into the core of Nature"—
O Philistine—
"No earthly mind can enter."
The maxim is fine;
But have the grace
To spare the dissenter,
Me and my kind.
We think: in every place
We're at the center.
"Happy the mortal creature
To whom she shows no more
Than the outer rind,"
For sixty years I've heard your sort announce.
It makes me swear, though quietly;
To myself a thousand times I say:
All things she grants, gladly and lavishly;
Nature has neither core
Nor outer rind,
Being all things at once.
It's yourself you should scrutinize to see
Whether you're center or periphery.

Johann Wolfgang von Goethe
(Translated by Michael Hamburger)

The Weather of the World

Now that the cameras zero in from space
To view the earth entire, we know the whole
Of the weather of the world, the atmosphere,
As though it were a great sensorium,
The vast enfolding cortex of the globe,
Containing contradictions, tempers, moods,
Able to be serene, gloomy or mad,
Liable to huge explosions, brooding in
Depressions over several thousand miles
In length and trailing tears in floods of sorrow
That drown the counties and the towns. What power
There is in feeling! We are witness to,
Enslaved beneath, the passions of a beast
Of water and air, a shaman shifting shape
At the mercy of his moods, trying to bend,
Maybe, but under pressure like to break.

His mind is our mind and the world's alike,
His smiles, his rages and aridities,
Reflect us large across the continents
And improvise our inwardness upon
The desert and the sea; we suffer him
As if he were the sufferer buried in
The self and hidden in heaven's indifference;
And like us he seeks balances that are
Inherently unstable to either hand;
The id, the superego, and the god
Of this world, the apparent devil of the will
Whom God has given power over us
Or cannot or else will not bring to heel,
Our nourisher and need, our sorrow and rage,
Reacting and reflecting on our lives
In windy eloquence and rainy light
As in the brilliant stillness of the sun.

Howard Nemerov

The Specialist

He no longer marvels at stars.
That's what.

Under the telescope's dome,
he crouches in a
concrete basement,
his head down,
decoding the graphs
a cosmic camera
drops in his lap.

In his own regional head
(it could be a dungeon)
he parses electrical impulses,
regroups the lines
(fudging a little)
into lakes and sandstorms,
peaks and tide wracks,
his whole territory
shut against trespassers.

Anne S. Perlman

Artificial Intelligence

to G.P.S.

Over the chessboard now,
Your Artificiality concludes
a final check; rests; broods—
no—sorts and stacks a file of memories,
while I
concede the victory, bow,
and slouch among my free associations.

You never had a mother,
let's say? no digital Gertrude
whom you'd as lief have seen
Kingless? So your White Queen
was just an "operator."
(My Red had incandescence,
ire, aura, flare,
and trapped me several moments in her stare.)

I'm sulking, clearly, in the great tradition
of human waste. Why not
dump the whole reeking snarl
and let you solve me once for all?
(*Parameter:* a black-faced Luddite
itching for ecstasies of sabotage.)

Still, when
they make you write your poems, later on,
who'd envy you, force-fed
on all those variorum
editions of our primitive endeavors,
those frozen pemmican language-rations
they'll cram you with? denied
our luxury of nausea, you
forget nothing, have no dreams.

Adrienne Rich

In Computers

In the magnets of computers will
  be stored

  Blend of sunset over wheat
    fields.
  Low thunder of gazelle.
  Light, sweet wind on high
    ground.
  Vacuum stillness spreading from
    a thick snowfall.

Men will sit in rooms
upon the smooth, scrubbed earth
or stand in tunnels on the moon
and instruct themselves in how it
  was.
Nothing will be lost.
Nothing will be lost.

Alan P. Lightman

Moon Landing

It's natural the Boys should whoop it up for
so huge a phallic triumph, an adventure
    it would not have occurred to women
    to think worth while, made possible only

because we like huddling in gangs and knowing
the exact time: yes, our sex may in fairness
    hurrah the deed, although the motives
    that primed it were somewhat less than *menschlich*.

A grand gesture. But what does it period?
What does it osse? We were always adroiter
    with objects than lives, and more facile
    at courage than kindness: from the moment

the first flint was flaked this landing was merely
a matter of time. But our selves, like Adam's,
    still don't fit us exactly, modern
    only in this—our lack of decorum.

Homer's heroes were certainly no braver
than our Trio, but more fortunate: Hector
    was excused the insult of having
    his valor covered by television.

Worth *going* to see? I can well believe it.
Worth *seeing?* Mneh! I once rode through a desert
    and was not charmed: give me a watered
    lively garden, remote from blatherers

about the New, the von Brauns and their ilk, where
on August mornings I can count the morning
    glories, where to die has a meaning,
    and no engine can shift my perspective.

Unsmudged, thank God, my Moon still queens the Heavens
as She ebbs and fulls, a Presence to glop at,
    Her Old Man, made of grit not protein,
    still visits my Austrian several

with His old detachment, and the old warnings
still have power to scare me: Hybris comes to
    an ugly finish, Irreverence
    is a greater oaf than Superstition.

Our apparatniks will continue making
the usual squalid mess called History:
    all we can pray for is that artists,
    chefs and saints may still appear to blithe it.

August 1969

W.H. Auden

172

Koko

They have given the gorilla language,
but not through the hopeless throat,
its larynx a fiercely clenched bulb
that never opened.

Instead they press her thick licorice digits
into mute nouns older than Babel,
ancient as bones in tar.
This is taxidermy on the living,
and charades with the dead.

So *gesticulate*
joins the lexicon of
masticate, *urinate*, and *masturbate*.
The gorilla cleans her trailer,
then tears up the sponge,
hosts a tea-party for the deaf
student who often visits,
offering refreshment in a toy cup,
eating a crayon as
daintily as if it were a meringue.

But when her trainer has to go
the tar-baby prodigy cries,
for all the proud sum of her vocabulary,
signing to herself in the dark
about white tigers,
not knowing a word for *zebra*.

Ann Downer

In Distress

(Selected entirely from International Code of Signals, *United States Edition, published by U.S. Naval Oceanographic Office)*

I am abandoning my vessel
Which has suffered a nuclear accident
And is a possible source of radiation danger.
> *You should abandon your vessel as quickly as possible.*
> *Your vessel will have to be abandoned.*

I shall abandon my vessel
Unless you will remain by me,
Ready to assist.
I have had a serious nuclear accident
And you should approach with caution.
The position of the accident is marked by flame.
The position of the accident is marked by wreckage.
I need a doctor. I have severe burns.
I need a doctor. I have radiation casualties.
I require a helicopter urgently, with a doctor.
The number of injured or dead is not yet known.
Your aircraft should endeavor to alight
Where a flag is waved or a light is shown.
Shall I train my searchlight nearly vertical
On a cloud intermittently and, if I see your aircraft,
Deflect the beam upwind and on the water
To facilitate your landing?
> *I do not see any light.*

You may alight on my deck; I am ready to receive you forward.
You may alight on my deck; I am ready to receive you amidship.
You may alight on my deck; I am ready to receive you aft.
> *I am entering a zone of restricted visibility.*
> *Visibility is decreasing.*
> *You should come within visual signal distance.*

I require immediate assistance; I have a dangerous list.
I require immediate assistance; I have damaged steering gear.
I require immediate assistance; I have a serious disturbance on board.
I require immediate assistance; I am on fire.
> *What assistance do you require?*
> *Can you proceed without assistance?*
> *Boats cannot be used because of weather conditions.*
> *Boats cannot be used on the starboard side because of list.*
> *Boats cannot be used on the port side because of list.*
> *Boats cannot be used to disembark people.*
> *Boats cannot be used to get alongside.*
> *Boats cannot be used to reach you.*
> *I cannot send a boat.*

I require immediate assistance; I am drifting.
I am breaking adrift. I have broken adrift.
I am sinking.
> *Did you see the vessel sink?*
> *Is it confirmed that the vessel has sunk?*
> *What is the depth of water where the vessel sank?*
> *Where did the vessel sink?*
> *I have lost sight of you.*

My position is doubtful.
My position is ascertained by dead reckoning.
Will you give me my position?
> *You should indicate your position by searchlight.*
> *You should indicate your position by smoke signal.*
> *You should indicate your position by rockets or flares.*

My position is marked by flame.
My position is marked by wreckage.
Are you in the search area?
> *I am in the search area.*

Are you continuing to search?
> *Do you want me to continue to search?*
> *I cannot continue to search.*

I cannot save my vessel.
Keep as close as possible.
I wish some persons taken off.
A skeleton crew will remain on board.
You should give immediate assistance to pick up survivors.
You should try to obtain from survivors all possible information.
> *I cannot take off persons.*
> *There are indications of an intense depression forming.*
> *The wind is expected to veer.*
> *You should take appropriate precautions.*
> *A phenomenal wave is expected.*
> *I cannot proceed to the rescue.*
> *I will keep close to you during the night.*
> *Nothing can be done until daylight.*

David Wagoner

The Cat in the Box

A cat is placed in a box.
Inside, a device can release a fatal gas.
A random atomic event
will determine
if the gas is released.

The box is closed.
A moment later, the gas
has either been released or not released.

> Acrid with fear and sweat,
> the people are jostled, prodded,
> herded to the showers, handed towel and soap.
> The door clanks shut, a technician
> moves the lever.

According to classical physics
the cat is either dead or not dead.
According to my mother, my dentist,
to common sense.

> The people lie piled atop one another,
> limp on the concrete floor.
> Unless the plumbing failed; in which case,
> they still await the water's stream.

Most quantum mechanicists
the hidden cat lies in
a netherland between the two possibilities
and the actualization of one.
Only when someone lifts the lid to look inside
does one possibility vanish.
Only then is the cat dead or alive.

If the guard is particularly meticulous
in how he lines up the next bunch,
those inside may hover longer, cloth
in hand, awaiting the steely gaze
to release them into death.

Those physicists who subscribe to
the Many Worlds Interpretation
claim that when the experiment begins,
the world splits into two,
each with a cat in a box.
The worker who peeks into the box
becomes two, each oblivious of the other.
One self sees a dead cat,
the other a live cat.

As in another world,
the guard opens the door to find
people impatient to wash. Perhaps a few
even dancing, joyful with the knowledge
of what they missed.

Ann Rae Jonas

Toward Climax

I

salt seas, mountains, deserts—
cell mandala holding water
nerve network linking toes and eyes
fins   legs   wings—
teeth, all-purpose little early mammal molars.
primate flat-foot
front fore-mounted eyes—

watching at the forest-grassland (interface
richness) edge.
scavenge, gather, rise up on rear legs.
running—grasping—hand and eye;
hunting.
calling others to the stalk, the drive.

note sharp points of split bone; broken rock.

brain-size blossoming
on the balance of the neck,
tough skin—good eyes—sharp ears—
move in bands.
milkweed fiber rolled out on the thigh;
      nets to carry fruits or meat.

catch fire, move on.
eurasia tundra reindeer herds
sewn hide clothing, mammoth-rib-framework tent.

Bison, bear,   skinned and split;
      opening animal chests and bellies, skulls,
      bodies just like ours—
pictures in caves.

send sound off the mouth and lips
formal complex grammars transect
      inner structures & the daily world—

big herds dwindle
    (—did we kill them?
    thousand-mile front of prairie fire—)
ice age warms up
learn more plants. netting, trapping, boats.
bow and arrow. dogs.
mingle bands and families in and out like language
    kin to grubs and trees and wolves

    dance and sing.
begin to go "beyond"—   reed flute—
    buried baby wrapped in many furs—
great dream-time tales to tell.

squash blossom in the garbage heap.
    start farming.
cows won't stay away, start herding.
weaving, throwing clay.
get better off, get class,
make lists, start writing down.

    forget wild plants, their virtues
    lose dream-time
    lose largest size of brain—

get safer, tighter, wrapped in,
winding smaller, spreading wider,
lay towns out in streets in rows,
and build a wall.

drain swamp for wet-rice grasses, burn back woods,
herd men like cows.
have slaves build a fleet

raid for wealth—bronze weapons
horse and wagon—iron—war.

study stars and figure central
never-moving Pole Star King.

II

From "King" project a Law. (Foxy self-
survival sense is Reason, since it "works")
and Reason gets ferocious as it goes for
order throughout nature—turns Law back on
nature. (A rooster was burned at the stake
for laying an egg. Unnatural. 1474.)

III

science walks in beauty:

nets are many knots
skin is border-guard, a pelt is borrowed warmth;
a bow is the flex of a limb in the wind
a giant downtown building
        is a creekbed stood on end.

detritus pathways. "delayed and complex ways
to pass the food through webs."

maturity. stop and think. draw on the mind's
stored richness. memory, dream, half-digested
image of your life. "detritus pathways"—feed
the many tiny things that feed an owl.
send heart boldly travelling,
on the heat of the dead & down.

IV

two logging songs

Clear-cut

Forestry. "How
Many people
Were harvested
In Viet-Nam?"

Clear-cut. "Some
Were children,
Some were over-ripe."

Virgin

A virgin
Forest
Is ancient; many-
Breasted,
Stable; at
Climax.

    Gary Snyder

Berceuse

Listen to Gieseking playing a Berceuse
of Chopin—the mothwing flutter
light as ash, perishable as burnt paper—

and sleep, now the furnaces of Auschwitz
are all out, and tourists go there.
The purest art has slept with turpitude,

we all pay taxes. Sleep. The day of waking
waits, cloned from the phoenix—
a thousand replicas in upright silos,

nurseries of the ultimate enterprise.
Decay will undo what it can, the rotten
fabric of our repose connives with doomsday.

Sleep on, scathed felicity. Sleep, rare
and perishable relic. Imagining's no shutter
against the absolute, incorrigible sunrise.

Amy Clampitt

Of How Scientists Are Often Ahead of Others in Thinking, While the Average Man Lags Behind; and How the Economist (Who Can Only Follow in the Footsteps of the Average Man Looking for Clues to the Future), Remains Thoroughly Out of It

Atomic Holocaust! Fall out! Consequences of Conflagration
    & Catastrophe! Thinking about the Unthinkable
    & The Day After. Not to mention Pollution

Of Stratosphere, Zoosphere, & Ionosphere! Mind-Blowing! We look
    under our books, radiator covers, soup-tureens, &
    nutshells, forgetting about our gloves. Where will
    the next war be fought?

Where will the next war be fought? Einstein pronounces:
    (ca. 1950) "Although I do not know the means
    by which, World War III will be fought, this much
    I can tell you: World War Four will be fought with

Sticks & Stones." Later (ca. 1990), teams of super-far-sighted
    scientists, in response to various urgent pleas (and
    after working together all night in a think-tank
    with only a suffocating goldfish in a bowl to look at)

Announce: "Never Fear.—Nobody else has thought of it
    but where the next war will be fought, is well-above
    the earth's atmosphere, in a positively sub-lunar

Safety. Some argument about air-space, betwixt the space-
    shuttles, may trigger it. When they explode
    the earth will not suffer the damage so much as
    the weight of a falling, splintered toothpick."

The Average Man in the 1980's, what has he been thinking?
    Shelters against fall-out—they're for the extra-
    urban, or the affluent. With his usual impeccable,
    lagging logic, the Average Man has gone out & bought
    an extra ten hats for his closet. & The Economist?

—The Economist has gone out & heavily invested in hat-stocks.

Michael Benedikt

The Fundamental Project of Technology

"A flash! A white flash sparkled!"

—*Tatsuichiro Akizuki,*
Concentric Circles of Death

Under glass, glass dishes which changed
in color; pieces of transformed beer bottles;
a household iron; bundles of wire become solid
lumps of iron; a pair of pliers; a ring of skull-
bone fused to the inside of a helmet; a pair of eyeglasses
taken off the eyes of an eyewitness, without glass,
which vanished, when a white flash sparkled.

An old man, possibly a soldier back then,
now reduced down to one who soon will die,
sucks at the cigarette dangling from his lip, peers
at the uniform, scorched, of some tiniest schoolboy,
sighs out bluish mists of his own ashes over
a pressed tin lunch box well crushed back then when
the word 'future' first learned, in a white flash, to jerk tears.

On the bridge outside, in navy black, a group
of schoolchildren line up, hold it, grin at a flash-pop,
scatter across grass, see one from elsewhere, cry,
hello! hello! hello! and soon, *goodbye! goodbye! goodbye!*
having pecked up the greetings that fell half unspoken
and the going-sayings that those who went the morning
it happened a white flash sparkled did not get to say.

186

If all a city's faces were to shrink back all at once
from their skulls, would a new sound come into existence,
audible above moans eaves extract from wind that smooths
the grass on graves; or raspings heart's-blood greases still;
or wails babies trill born already skillful at the grandpa's rattle;
of infra-screams bitter-knowledge's speechlessness
memorized, at that white flash, inside closed-forever mouths?

To de-animalize human mentality, to purge it of obsolete
evolutionary characteristics, in particular death,
which foreknowledge terrorizes the contents of skulls with,
is the fundamental project of technology; however,
the mechanisms of *pseudologica fantastica* require:
to establish deathlessness it is necessary to eliminate
those who die; a task attempted, when a white flash sparkled.

Unlike the trees of home, which continually evaporate
along the skyline, the trees here have been enticed down
toward world-eternity. No one knows which gods they enshrine.
Does it matter? Awareness of ignorance is as devout
as knowledge of knowledge. Or more so. Even though not knowing,
sometimes we weep, from surplus of gratitude, even though knowing,
twice already on earth sparkled a flash, a white flash.

The children go away. By nature they do. And by memory:
in scorched uniforms, holding tiny crushed-metal lunch tins.
All the shinto-groans of each night call them back, satori
their ghostliness back into the ashes, in the momentary shrines,
the thankfulness of arms, from which they will go
again and again, until the day flashes and no one lives
to look back and say, a flash, a white flash sparkled.

Galway Kinnell

Call to Order

And the order of the universe is reversed.

And the reversed order of the universe is
Unisexual, unlikely, unilateral, uninterested.

What we have left behind,
Our empty insect shells in the sand,
As if we had evolved
And wandered away from our parasol,
Will become sand,
The reversed order of flesh and trees,

The referee of trees, the wind the ball
Speeding against the spaces in the day.
The world, when we are gone, will be a bone
Pocked in the gape of the morning.

It will sound like fish.
It will swim in silence without thinking
What the order, what the reversed order,
What the reversed order of the universe
Is, what it is to be.

If the sky is blue,
Will it miss our trained eyelids?
Would the air prefer to carry our voices
In the old way?

Where we will go
Is the expectation of death,
Again and again no sound.
We arrive full of the recognized,
Fruit must be fruit still;
Ribbons, glass, feet, fingernails,
But none of it will be there.
It will not be there.

Carol Burbank

Toward a Theory of Instruction

I

Like many hard-shelled, luckless creatures
of this world, the hermit crab
mates but once a year. Imagine,
if you will, a tide-line roiling with
their rat-tails thrashing, the violent
ebb and flow of aphrodisiac juices
and sperm frothing the surf,
twenty males, count them, for every
female; they look like a pile of World
War I helmets. What's going on here?

High in the Sierra Madre mountains
there is a species of insects that mate
several times a day, often up to
three dozen. Of course, these bugs
are bisexual and they all look alike,
so scientific observations and data
are somewhat unconfirmed. In fact,
word of these insects' existence
is mere hearsay. In fact, I made
all this up. What's going on here?

I woke today from a nap
to the loudest sound I've ever heard.
I don't know if I was dreaming.
Perhaps I heard the world exploding.
Perhaps I had a stroke.
No one admitted to dropping anything.
I looked out to snow swirling
in the early evening streetlight.
I remember only one previous scare
that shook me as much. I dreamed
I was in a building that was collapsing.

What's going on here, I thought,
then and today. One lone
snowflake touched the window
and I heard the sound again.

II

Possibilities for a new world:

1. It will be all covered with hair.

2. In the clam-colored light through the winter windows,
your long arms on the table, your round arms on the table,
your arms the color of the moon covered with a fine simian
hair, the resonant architecture of your hands creating
small realities before my eyes, small talk, glittering beer
sculpted by transparency of—what?—our daily, raging lives,
whatever's going on in the breathing, earnest air between us
in the clam-light, the moonlight of your skin. What's
going on here?

How to explain the behavior of anything?
We hack at each other with the largest
knives we can find. Animals, in their simplicity,
do everything better. In ignorance.
In ignorance, we sputter out like wet matches.
We comb our hair wrong. We don't talk. Or don't
listen. We ravage each other's hearts like
eggplants. We sacrifice each other like 25 dollar
cords of firewood.

Or 3. It is a movie from the 1930's.
It is raining. Lights reflect off the shining street.
In the doorway of a very old brick building sleeps
a very old man. He is dreaming of the third
possibility for a new world. It will be a world
playing with a full deck, a world with both oars
in the water. A world where beauty runs
in the meadow all day long. The world will be
the old man's oyster, the apple of his eye.

Meanwhile, a sharpie and his girlfriend run by him
on their way to the movies. They don't care
what's playing. They want to hold hands
and kill some time. Until the rain moves on.

III

Hermit crabs mate on the longest day
of the year. And they take their time.
Some of them take so much time, the tide abandons them
on the beach. So they must wait twelve hours
for the next high tide. They don't mind.
They merely burrow down into the wet sand.
And have another cigarette. And talk about
growing up in New England.

Danny Rendleman

Family Chronicle

Vegetation in the watershed
recrystallization in the highlands
during the years of bloated animals—years
when the children were born dead
or with cataracts & extra fingers.

There was a drone of rosaries at the novenas
gypsies crowded into the presbyteries
&
meteorites
deteriorated in museums.

It was blamed on low energy galactic particles
or the cooling solar nebula; crustal
heterogeneities—quiet fluxes of ions.

In the grottos of dead membranes
microscopic spherules, ellipsoides, & shards
created a glacial silt
an immunological paralysis.

The bones in our mausoleum
were sifted through for extra room:
seminarians, gardeners, & coachmen
grandmothers, & spinster aunts
& all those babies.

Anselm Parlatore

Advice to a Prophet

When you come, as you soon must, to the streets of our city,
Mad-eyed from stating the obvious,
Not proclaiming our fall but begging us
In God's name to have self-pity,

Spare us all word of the weapons, their force and range,
The long numbers that rocket the mind;
Our slow, unreckoning hearts will be left behind,
Unable to fear what is too strange.

Nor shall you scare us with talk of the death of the race.
How should we dream of this place without us?—
The sun mere fire, the leaves untroubled about us,
A stone look on the stone's face?

Speak of the world's own change. Though we cannot conceive
Of an undreamt thing, we know to our cost
How the dreamt cloud crumbles, the vines are blackened
    by frost,
How the view alters. We could believe,

If you told us so, that the white-tailed deer will slip
Into perfect shade, grown perfectly shy,
The lark avoid the reaches of our eye,
The jack-pine lose its knuckled grip

On the cold ledge, and every torrent burn
As Xanthus once, its gliding trout
Stunned in a twinkling. What should we be without
The dolphin's arc, the dove's return,

These things in which we have seen ourselves and spoken?
Ask us, prophet, how we shall call
Our natures forth when that live tongue is all
Dispelled, that glass obscured or broken

In which we have said the rose of our love and the clean
Horse of our courage, in which beheld
The singing locust of the soul unshelled,
And all we mean or wish to mean.

Ask us, ask us whether with the worldless rose
Our hearts shall fail us; come demanding
Whether there shall be lofty or long standing
When the bronze annals of the oak-tree close.

Richard Wilbur

Magnificat in Transit from the Toledo Airport

The world has a glass center.
I saw the sign for it.

TOLEDO, GLASS CENTER OF THE WORLD

That's what surprised me.
I mean that it was Toledo.
I knew the center was glass.
That's why we've got this cleaning-and-polishing operation going.
There were bulldozers outside of Toledo, working away.

It's a beginning.
Move this junk, we'll be able to see in.
When the Chinamen at the other end gear up, we'll be able to see
    through.
It's like the completion of the first transcontinental railroad.

It's like what the Egyptians had in mind when they invented the
    pyramids.
It's like what God would have opted for if he had been an optician.
There it'll be.
This really tremendous lens.

Think of the excitement when they put the ceremonial white
    handkerchief
into the outstretched hand of the Final Polisher.
What if he suddenly thinks he's Sonja Henie?
What if he just gets awestruck and sits down?

Actually there'll be an airtight operating procedure.
Polish it off. Take measurements. Melt in.

You can do that, when it's glass.
Goes into what they call a "solid solution."
"Doping the mix," they call it. Vary the additive
and the whole ball of glass comes out in a wonderful color:
rose. ultramarine. turquoise. maroon.
It's enough to make you lose your marbles.

It *is* a marble.

I know all about marbles, I said, this is my bag!
Will it be a milkie? Will it be a purie? Will it be a swirl?
"Oh get more bulldozers, Ohio," I said to myself in a fit of eloquence.
"Toledo *is* the glass center of the world."

I felt like the *Boy's Life* monthly good luck stories.
"Cabin Boy with Columbus." "Young Hank Ford."

I always liked the one where the safari
saves me (I've just been orphaned by the Zulu)
and lets me tag along with them up country
into the wild unknown Witwatersrand.
You know the story: how the evil porter
filches the beads, and fills the bags with stones.

Toledo Toledo, I warbled,
even though this Red Yellow Cab Company taxicab is bearing me away
        from you at maximum achievable velocity,
even though my appointment is in Bowling Green,
I shall push in the press for deployment of Project Ploughshare.
I shall lobby. I shall make mailings. I shall crusade.
Instant redistribution of silage among the developed peoples!
Riverbanks full of afterburgers! Boom!
Is there a high-gain, high-risk, maximum-impact slot in there somewhere
        for a creative projects person with hands-on image-management
        experience and a desire to position himself on the forefront of
        America's growth industry for the 1990s?
Toledo, I have arrived!

I have arrived in Bowling Green.

Bowling Green State, the Athens of the Midwest.

Emptiness. Houses. Emptiness-houses. Shrublets.
Grass. What am I doing here? This trike.

Here, from his door, to my rescue, comes my host.
He is Ray DiPalma.
He is a poet.
He is from Pittsburgh, Steel Center of the World.

Oh damn. Wouldn't you know? He is. It's true.

He is from Pittsburgh Steel Center of the World
And *I* knew that. *I* used to have a steelie.
Too heavy, they said, that's cheating, you go home.

You ever look down deep into a steelie?
Really strange reflections.
Whole horizon.
You! bigger than everybody's houses.
Big-nose.
Steely little eyes.

You know what's wrong with American free enterprise?
It's two-bit people in two-bit little places
doin' a two-bit number on theirselves.

Myself I know the world has a glass center.
Myself I know it's Toledo, built on sand.
And soon now, in the place of the bulldozer,
soon now, under the battlements and banners,
rose, ultramarine, turquoise, maroon,
this scraggy ol', scruffy ol', vermin-infested slagpile,
this rubble-in-arms, gonna sing out loud an' clear,
like a great bell, like the choir of all the angles.
I'll get more dope on it as soon as I get home to Iowa City.

Iowa City — the Athens of the Midwest.
I have a friend there in the Pittsburgh Plate Glass Company.
Damned if it isn't true — Pittsburgh Plate Glass.

Not Pittsburgh Paint now, mind you. Pittsburgh Paint
sells glass put out by Owens Illinois.
At least that's what it does in Iowa.
Or did, when I last had a broken window.
To tell the truth, that was a while ago.
Recently I've been living in New Hampshire.

They make a claim there about maple syrup
but I say let's pretend they never said it.

I mean I'm all for backwoods boosterism
but when you start to ask folks to believe
this mighty item all wrapped up in blue,
this whopping gobbet of long-lasting goodness,
this great big hunk, this Earth, this greater Mars,
is something with a novelty interior —
a sort of oversized designer chocolate —
well that's beyond preposterous. That's dumb.

Sometimes when I consider how this world
is given to outright exaggeration,
I think I should've stood put in Columbus.

Columbus Ohio, the Athens of the Midwest.
Home of the mighty Buckeyes. I was born there
and moved away when I was four days old
and always like to tell it that way, acting
as if I'd sized the place up and skedaddled.

The Buckeyes are Ohio State of course.
Ohio State is not Ohio U.
Ohio U's a college, and a good one,
off in a little mill town name of Athens.
God knows what Athens is the Athens of.

Actual buckeyes — just in case you wondered —
Actual buckeyes are the kind of nut
you'd pick up off the sidewalks in October
and ram a toothpick into for a spindle
to hold a set of cut-out paper sails
and join Miss Dean's Columbus-Day flotilla
across the blue construction-paper ocean
to prove the world was chock-a-block with marvels
to *make* things out of, Right Beneath Your Feet.

I think she made me a Discoverer,
Miss Dean did. And Miss Whitman. And Miss Ide.
I got the real good news about creation.

Merciful Muchness, Principle of Redundance,
Light of the World converted at every turn,
the world *is* too much for us, wait and see!

George Starbuck

Dr. Dimity Lectures on Unusual Cases

The laparotomy revealed a tumor of the insulin-secreting
Cells of the pancreas. Once the offending tumor had been
Removed, her behavior returned to normal and she was
Once again a dutiful, conscientious wife and mother.

The next case was much more difficult to diagnose—impossible
       Three cardiologists before me said,
Three well-known cardiologists, that is. They knew the heart
       Had been affected.
Well, any third-year student could have ascertained as much.
       Even the parents could see their daughter
Turning bluer every day, like litmus paper when the moisture level
       Changes. But all the films and graphs
And probes did not reveal the cause. I do not know what made me
       Place my head against her chest exactly
As a doctor would have eighty years ago. But it is those unexpected
       Bursts of inspiration which separate
The competent from the superior. My head against her breast.
       It heard the faintest scratching,
Quite different from the rubs or murmurs we are trained to hear.
       I strained to match sound with memory for
Several days. Oblivious to almost everything, I wore a sport shirt
       To the office and asked
For "sandwich" instead of "scalpel" in a mitral valve replacement.
       Then, in the kitchen on my third
Sleepless night, I got it or, rather, heard it; three scurried away
       When I turned on the light.
She is alive and well, now, for it was simple to remove once
       I had identified the problem.
Which orifice it entered through or how it lived within
       A liquid medium we do not know.
Those are aspects which a class on diagnosis need not touch.

Next week I will discuss:
Life and the Liver in Sibling Kidney Transplant.

Cynthia Macdonald

Dr. Dimity Is Forced to Complain

Dr. Dimity's head hurts. He refuses to use a stronger word.
He lies in his darkened bedroom, groaning, trying
To contain his pain. The smell of soup
Which permeates the house, even though his wife has
Turned off the stove, makes him feel worse.
He asks her to send for Dr. Doctor who cancels
His next two appointments and postpones his golf date.
Dr. Dimity, his friend and colleague, never complains.

Dr. Doctor begins with the eyes and needs look no further.
Through his ophthalmoscope, instead of the usual vascular pattern
Radiating from the optic nerve, he sees straight
Horizontal lines of light. "I see
Straight horizontal lines of light," he tells Dr. Dimity
Who groans, trying not to shut his eyes.

"Wait, the lines are widening. There are slats between them.
It is a Venetian blind. Being pulled up. I am not sure
If I am looking out or looking in."
A storm, trees bending, breaking.
The sky is being ripped by wind. A house.
Grounds, bushes under water. It is moving fast,
Rising fast. Debris batters the house.
A man and a boy lean out the window, looking
At the rising water. The boy is crying.
Dr. Dimity shuts his eyes, "That is enough."

"It was like a nickelodeon," says Dr. Doctor.
"I would write this up for *Internal Medicine* if
I could make a diagnosis. I know you will agree physicians
Present the most baffling problems and
Are the least cooperative."
"A form of migraine," says Dr. Dimity.
"A flood of painful memories," says Dr. Doctor.

Cynthia Macdonald

Chicken Soup Therapy: Its Mode of Action

A recent issue of *Chest*[1] contains an interesting report of a patient who, having received an inadequate course of chicken soup for the treatment of a mild pneumococcal pneumonia, developed a severe relapse. Although the pharmakinetics of such therapy are mentioned, its mechanism of action is not delineated:

> Chicken soup as therapy for many ills
> Has a much more honored place than pot or pills.
> Years and trials in the field have proved its use,
> Though some workers[2] study now its drug abuse.
> But its mode of action is not widely known:
> We have found that when it's boiled with no bone
> Potency and clinical response are less.
> Also, organ distillates seem to possess
> Factors that allow the optimal effects.
> These results suggest that chicken soup protects
> By providing marrow cells and factors for
> Passive host immune defense; But even more,
> Transferred from man's fowl friend are Bursal Brews
> Sought so long by modern medicine's gurus.

CAROLINE BREESE HALL, MD
Infectious Disease Unit
University of Rochester Medical Center
260 Crittenden Blvd
Rochester, NY 14642

References

1. Caroline NL, Schwartz H: Chicken soup rebound and relapse of pneumonia: Report of a case. *Chest* 67:215–216, 1975.
2. Campbell, Heinz: *A & P*, 1975.

Plato Instructs a Midwest Farmer

"It is in the fake picture that we first perceive
reality, in the contrived colors we find
light, in the stilted postures
relationships become clear for the first time.
The structure of the room in your painting above the fireplace
which exists only in the painter's mind
is more real, more present than the room
in which we sit talking. We see only shadows
and reflections. Put another way
it is true that Bach, Haydn, and Mozart
record the basic sound, the ultimate harmony
the final Form of the universe
and the jangling tractor outside your window
is a perversion of sound, a kind of rudeness
an intrusion, not true, not complete.
The tractor's crude rhythm would rise toward a Bach chorale."
"Well," replied the farmer, kicking a stone
protruding from the fireplace until his toe hurt
"I don't think it's gonna make it."

David Palmer

Teeth

Teeth are a rather ridiculous inside remnant of the outside. Their life is filled with dread that they'll be forced outside again and lost there. A lost tooth doesn't know if it's clenched or revealed in a smile, it doesn't know how to put down roots and so it loses its capacity for aching.

Many teeth have been lost throughout the history of civilization, some by educational and corrective measures in the lives of younger individuals, some by the wasting away of old age. Teeth that have been bashed in during the development of civilizations don't rot, they scurry along in the darkness, scared of daylight. Some just grow tired, and are discovered and described in some new scientific discipline.

Surviving teeth convene on dark cloudy evenings, trembling with horror and telling the old gum stories, fist stories, stories of stiff boots, of other kinds of bashing. These stories are not without some unintentional comic effects, the teeth become comic figures, repeating their plots evening after evening, century after century.

That's how the puppet theater came about.

It's the theater of teeth, for which there's no mouth.

Now close your mouths, children, and listen.

Miroslav Holub

St. Augustine Contemplating the Bust of Einstein

I

Selfless now in this crucible of light,
I act not by instinct, but rightly.
It's self-consciousness
that makes cowards of us all.
But the cloisters of my memory still
ring with visions of the concrete world,
that plane of one/many
riddling my life since *ex vacuo*.
Like you in my teens — a woolly
dream-gatherer — I worshiped the creature
over the Creator, prayed for
"chastity and continence, but not yet,"
then chained my senses together
like a great long molecule
blazing in some hyperbolic present.
I thought God dwelt in the motion of birds
rooting idly through damp sod,
or in their voluntary nervous system
of ducking their heads as they strut
in the cold. Skylocked, indigo branches
moved with more cadence than a Christian hymn.

II

But soon I became a house divided
between that geyser of color
and the asylum of God, *Fountain of Life*
and *Physician of my Soul*. Like you,
I blinkered my path to the essential,
ignored all the frippery that clutters up
the mind. You say photons clamor
eternally, and that speaks to me:
my first vision was one of infinite light.

III

But your principle of relativity
is as old as Galileo, who tried
to measure the speed of light
with two lanterns.
Your lanterns are planets,
and you conjure with such names
as One-Zwicky-One, Cygnus, and Orion,
prove your theorems with hydrofoils
that run on lubricated air.
I hear you've even computed
the circumference of the cosmos
(some exponential nightmare
reduced to ½ *a googol*),
but your gospel is no different
from Galileo's, only bolder:
you want to drive a stake
under the nail of the universe
and draw God out
like a soft-shelled crab.

IV

How elemental, the way you put it:
In the beginning
was the velocity of light.
But it means that God will never
be the same, now he's plummeted a peg
and had his secrets ogled.
You call this abomination Symmetry,
and that eludes me (light can behave
like a bullet or like the sound
of the shot, or like doubt
cruising silent as a white shark),
but when you say all is mutable,
that inflames me, because change
is contrary to the nature of God.
Where does that leave us?
Opting to face God or to face the Truth?
Or to face a God whose Truths are on tap,
some Holy Victim squatting on a pulsar?

Standing here bolt-still
(which is only to say: moving uniformly
at zero speed), I wonder if your myth
is right or just pretty, a kind of brainstem
sonata for μ-meson and cube root.
Tongue-tied, hamstrung, I feel battered
by answers that won't arrive for decades;
when they do, I suspect I'll find you
on the scaffold of the Knowable,
furiously soaping the noose.

Diane Ackerman

The Swarm

Blowflies explode from nowhere as I walk.
The hot pavement below me glares like chalk.
    Brown on the bleached cement,
I spot the bit of filth from which they flew,
crusted and dry, and poke it with my shoe,
        and it shines wetly. Intent
        buzzings return, bent
        in spirals on the food,
or quick scribbles of shrinking amplitude,

to the fruit, now impossible to find
under its fly-thick, greenly nervous rind.
        I watch them, half afraid.
It's breath-taking. They all crawl and conform.
Such eager industry! The civil swarm
        elegantly displayed.
        Their brightly chained brocade
        of green and golden links
covers a core of rottenness that stinks.

What social sense enables them to meet
on tiny morsels in this desert street
        and settle and belong?
I think they fly on instruments of smell.
Does anything that men do work so well?
        The missile might go wrong
        that fires and feels along
        its calculated flight
to the warm target, swarming in the night.

Yet though men blunder, they can make correction.
O Lord, thy servants labor toward perfection.
        How can they be like us,
these nudging bodies, shiny, tough, and quick?

My flesh shivers, crawls, feeling almost sick.
    Their innards are like pus.
    . What if I'm envious?
    Didn't I hope to shine,
pick, glut, in my great country's rapt design?

Even the poet, fishing for renown
and having found it, having played the clown,
    will call his labors blessed,
accept respect, drop his polite defiance,
and praise his country and the works of science—
    and spiral down to rest
    and comfort on his nest.
    O flesh, you're like these flies.
Does no one spiral into empty skies?

One leaves a wife maybe, or quits a job,
flies to a foreign country. . . . Still the blob
    called love and comfort waits—
only to rot when one has lost one's wit
and fortitude and spiraled down to it,
    pulled by the nasty fates.
    Then more and more one hates . . .
    one tires. The sinews twist.
All men grow ugly now. They still exist.

I stamp at them. The flies spiral about
and settle still. No one will stamp them out.
    Give me another sight!
In the next block, clustered on maple trees,
leaves in green swarms are crawling in the breeze. . . .
    At telescopic height
    in far corners of night
    science's watcher sees
fly-spirals mimicked in the galaxies.

                          Richard Moore

The Future

The future is withdrawn again.

Just as the donkey's teeth are pressing against the carrot, somebody
removes the donkey.

If this is all the future holds for our nation's donkeys we may never
make it through the forthcoming year!

What we need is a future in which vegetable gardens are planted in
the atmosphere.

Gnash gnash.

Each step of the way must be marked by a half-eaten tuber or legume
tossed aside, as if luxury were everyday

Or, as if donkeys could now go in any direction they wanted, even into
the past.

Advancing on all sides, listen to the delicate steppings and droppings
of the donkeys, catching up to the future, making the past come
alive sooner and sooner.

When the future is finally achieved, the valley of the world will see,
strewn as far as the eye can see, donkeys sitting down or lying
around at their ease, little carts unhitched and parked; we will be
aware as never before of the portability of time

Michael Benedikt

209

BIOGRAPHICAL NOTES

DANNIE ABSE, born in Cardiff, England in 1923, is a doctor, poet, novelist and playwright. With a wide English audience and several volumes of poetry, he has gradually acquired an American following. His work covers many themes, though he consistently returns to medical topics as he has in the poem printed here. (Page 76)

DIANE ACKERMAN was born in 1948 in Waukegan, Illinois, and studied writing at Cornell University where she became interested in using the details of science and technology in her work. Her first book, *The Planets: A Cosmic Pastoral,* was the result of a year's immersion in planetary astronomy while on a Rockefeller Graduate Fellowship in Humanities, Science, and Technology. Ackerman is currently teaching writing at Washington University in St. Louis, Missouri. She has recently become a pilot and her poems reflect this interest. (Pages 31, 42, 52, 204)

A.R. AMMONS is the Goldwin Smith Professor of Poetry at Cornell University. His longstanding interest in science has influenced at least a generation of students, several of whom also appear in this collection. Ammons' poetry has been honored by a Guggenheim Fellowship, the National Book Award, the National Book Critics Circle Award in Poetry, and a MacArthur Prize Fellow Award. (Pages 49, 122, 153)

W.H. AUDEN, 1907–1973, the English-born American writer, was one of the most important poets of the twentieth century. (Pages 50, 59, 67, 171)

LOIS BASSEN, who was educated at Vassar College and the City University of New York, was recently awarded a fellowship by the Mary Roberts Rinehart Foundation to complete a novel about mother and daughter scientists, *The Mother of Beauty.* She lives on Long Island with her husband and children, and teaches high-school and college English courses. Her poetry has appeared in national magazines, including *Science 84* and *The Kenyon Review.* (Page 123)

MARVIN BELL, born in 1937 in New York City, grew up in Long Island, New York. He currently teaches in the graduate writing program at the University of Iowa. He has published several volumes of poems and won national poetry awards, including the Lamont Award of the Academy of American Poets. (Page 78)

MICHAEL BENEDIKT'S professional, literary, and academic credits include five volumes of poetry. He has also edited two anthologies, one on the prose poem and one on the poetry of surrealism. He has lectured widely on literature and has taught at Vassar College, Bennington College, Sarah Lawrence College, Hampshire College, and Boston University. He currently lives in New York City. (Pages 85, 184, 209)

JODY BOLZ was born in 1949 in Washington, D.C. At Cornell University she studied with A.R. Ammons and received the Academy of American Poets College Prize. *Poems* by Diane Ackerman, Jody Bolz, and Nancy Steele was published by Stone-Marrow Press in 1973. Her work, often influenced by an interest in the natural world, has appeared in several magazines and anthologies. She is a magazine editor for The Nature Conservancy and teaches writing classes at George Washington University. (Page 141)

OLGA BROUMAS was born in Greece and trained in architecture and the graphic arts. She is the author of four poetry books, including *Beginning With 0*, for which she won the Yale Younger Poets Award in 1976. She teaches at Freehand College in Provincetown, Massachusetts, where she is using healing and bodywork to expand the capacity for ecstasy, memory, and expression. (Page 134)

CAROL BURBANK is a recent graduate of Boston University's Writing Program, where she studied with George Starbuck. She now teaches writing at Mitchell College in New London, Connecticut. Her work has appeared in *Rolling Stone Magazine* and literary publications. (Pages 75, 188)

MICHAEL CADNUM lives in Albany, California, and is currently a fellow in creative writing with the National Endowment for the Arts. He has published one book of poetry, *The Morning of the Massacre*. His poems have appeared in several magazines, including *Commonweal, Virginia Quarterly Review* and *Antioch Review*. (Page 62)

SIV CEDERING has published eight collections of poetry, written in English, and two novels and a book for children, written in Swedish. A new collection of poetry called *The Floating World* was printed by the University of Pittsburgh Press last fall. She currently lives in Amagansett, New York, where she is working on her first English-language novel and a book for children, plus poetry and drawings. In recent years Cedering has written many poems on astronomy and astronomers. "In my attempt to know the world around me, I continue to study both science and myth," says Cedering. "The name and specifics of a flower or a star are equally important to me; so is primitive or contemporary man's concept of the relationship of man to that flower and star. The intent is simply to discuss what it is to be human, at this time, in a universe that keeps on unfolding for the questioning mind." (Pages 10, 12, 14)

AMY CLAMPITT was born in Iowa and educated at Grinnell College. She lives in New York City, where she has worked as a nonfiction book editor and a reference librarian for the National Audubon Society. Her first full-length poetry collection, *The Kingfisher,* was nominated for the National Book Critics Circle Award in poetry in 1982. She has received a Guggenheim fellowship, and her poems appear in such magazines as *Poetry, The New Yorker, The American Scholar* and *The New Republic.* (Pages 102, 103, 112, 152, 183)

MICHAEL COLLIER was born in Phoenix, Arizona, in 1953 and educated at Connecticut College and The University of Arizona. He has received fellowships from the Fine Arts Center in Provincetown, the National Endowment for the Arts, and the Thomas Watson Foundation. In 1981, he won the *Discovery/The Nation* award for poetry. He lives in Baltimore and works as the director of poetry programs at the Folger Shakespeare Library in Washington, D.C. His poetry has appeared in *The Nation, Poetry* and *Poetry Northwest.* (Page 79)

e.e. cummings, 1894–1962, was a poet whose unusual use of typography, punctuation, and vocabulary was influential in the development of modern poetry. He also painted, and wrote novels and plays. (Page 163)

ANN DARR was born in Iowa and educated at the University of Iowa. She served as a pilot in the Army Airforce during World War II and worked as a radio script-writer in New York. She currently teaches writing classes at American University in Washington, D.C. and at the Writers Center in Maryland. Her awards include a *Discovery 70* prize from the Poetry Center in New York, a National Endowment for the Arts fellowship, and a Bunting Fellowship. Four collections of her poetry have been published; the most recent is *Riding with the Fireworks,* 1981. (Page 124)

LUCILLE DAY is a California poet who frequently draws upon her other professional concerns: neurochemistry and biology. The poems published in this collection are taken from her first book *Self-Portrait with Hand Microscope,* published in 1982. Her poetry has appeared in magazines including the *New York Times Magazine* and the *Hudson Review.* "It seems natural to me to use science in poetry because I've always been drawn to both, and have studied and practiced them simultaneously," Lucille writes. "In both science and poetry, I think we are constantly refining our view of the world, and finding out things are very different from what we thought we knew." (Pages 73, 74, 132)

EMILY DICKINSON, 1830–1886, is considered one of the finest and most original of American poets of the nineteenth century. Although she published only seven poems during her lifetime, she wrote hundreds. (Page 165)

PATRIC DICKINSON is a widely published British poet who lives near London. (Page 41)

ANNIE DILLARD'S best-known writing about the natural world is undoubtedly the prose book called *Pilgrim at Tinker Creek,* which won a Pulitzer in 1974. She has also written a substantial amount of poetry about science and natural history. She grew up in Pittsburgh, Pennsylvania, attended Hollins College, and currently lives in Middletown, Connecticut, where she teaches at Connecticut Wesleyan University. Dillard is especially fond of old natural-history books, from which she draws many of her ideas for poems. "I love their tone, their enthusiasm, their total innocence of moral questions, their pure love of the kind of childish material knowledge that drew so many of us to science when we were young," she says. "For me they recapture the innocence of science." (Pages 29, 87)

R.H.W. DILLARD lives in the Roanoke Valley of Virginia, where he directs the writing program at Hollins College. His most recent poetry collection is *The First Man on the Sun,* 1983. (Pages 15, 17)

JOHN DONNE, 1572–1631, was the most outstanding of the English metaphysical poets. He was also a churchman famous for his spellbinding sermons. (Page 147)

ANN DOWNER lives in Alexandria, Virginia, attended Smith College, and writes both poetry and short stories. (Page 173)

HELEN EHRLICH was born in Lorain, Ohio, in 1924. She has lived in Phoenix, Arizona, for 16 years. She entered college in 1976 and began writing poetry the following year. She has published poetry in national and local journals and this year her poem "Two Sonnets," which appeared first in *Science 82,* has been nominated for the 1983 Rhysling Award, sponsored by the Science Fiction Poetry Association. (Page 64)

LOREN EISELEY, 1907–1977, was an American anthropologist and author of eloquent books on science and human nature. He also wrote poetry. (Pages 109, 110)

LAURA FARGAS is a Washington, D.C. poet whose work has appeared in *Science 83* and national literary magazines. She was a finalist in the *Discovery/The Nation* poetry contest in 1983 and gives poetry readings whenever possible in the mid-Atlantic region. (Pages 47, 144, 157)

CARROL FLEMING was born and raised in California. Though she spent the last six years sailing in the Carribbean, she once again lives in San Diego with her husband and very young daughter. Her poems have appeared in national publications, including *Science 83,* and in anthologies. She is also a nonfiction writer whose stories and reviews have appeared in *Smithsonian, Science 84, Americas* and several major newspapers. One of her chief interests, and the subject of many stories and poems, is in observation of the natural world. (Page 126)

214

ROBERT FRAZIER has published about two hundred poems, most of them with scientific themes. In the last three years his work has shifted from science fiction to hard science. His work includes a long-running series of biographical cameos of scientists for *Isaac Asimov's Science Fiction Magazine*. He has published one collection, *Peregrine,* from Salt Works Press and has edited a poetry anthology on science fiction and scientific poems. Frazier says he is "continually drawn to science in his work because the majority of new words in the English language, and the majority of new concepts and viewpoints are derived from the sciences." (Pages 9, 71, 81)

MARTIN GALVIN writes poetry and short stories and has published widely in national magazines, including *Commonweal, Texas Review* and *Science 84.* He teaches English literature in schools in the Washington, D.C. area and teaches writing part time at the Writer's Center and Northern Virginia Community College. "Science seems a natural place to find wonderous facts and bedrock metaphor," says Galvin. "I feel a real affinity with and for that kind of combination of fact and fancy that the scientist must cultivate." (Page 162)

JOHANN WOLFGANG VON GOETHE, 1749–1832, is recognized as one of the greatest and most versatile European writers and thinkers of modern times. A German poet, playwright and novelist, he profoundly influenced the growth of literary romanticism. (Pages 84, 166)

CAROLIN BREESE HALL is a pediatrician and infectious-disease researcher at the University of Rochester Medical Center. She is married to the Chief of Medicine of Rochester General Hospital, William J. Hall, and has three children. She has written several books on pediatrics and on infectious disease, and published widely, including some poetry, in medical journals. (Page 201)

JOHN HAINES has published two books of poetry, *News From The Glacier* and *Living Off The Country,* which include several poems of scientific theme. He lives and works in Fairbanks, Alaska. (Page 131)

ROALD HOFFMANN was born in Galicia, Poland, just before World War II. He came to the United States in 1949 and was educated in New York public schools before entering Columbia University. The recent Nobel Prize recipient is an applied theoretical chemist at Cornell University and uses quantum mechanics to describe the motions of electrons in molecules. "The humanities, especially literature and the visual arts, have always been close to me," he says. "I see poetry as another way of trying to understand the universe, complementary to science ... I wish to write poetry about everyday science, an activity of man like any other, and I wish to use scientific images naturally." (Pages 80, 120)

MIROSLAV HOLUB, who has been called the most important poet working in Europe today, is chief research immunologist at the Institute for Clinical and Experimental Medicine in Prague. He writes about many subjects, but integrates scientific thought into his poetry. He hopes that "logical, analytical thinking about the human condition, and not abstract idealism, will result." Holub occasionally visits the United States to lecture and read his poetry, and he has taught at Oberlin College. Several of his thirteen books of poetry have been translated into English. (Pages 4, 5, 63, 90, 105, 106, 156, 161, 203)

ANNA RAE JONAS is a poet and nonfiction writer from New York City. After graduating from Goddard College in 1970, she supported her writing as a waitress, printer, architectural drafter and tour guide. In her thirties she became interested in scientific ideas and began using them in her poetry. She has been studying science writing with Horace Freedland Judson and earned an M.A. from Johns Hopkins University in June 1984. (Pages 138, 177)

JOSIE KEARNS has college degrees in English literature and in psychology. She works in the Chemistry and Resource Science Departments at the University of Michigan. She has written and published poetry widely and has won regional prizes for her work. "Because of the emphasis on design in science, science merges easily with poetry and also art in today's world. Science and technology have overwhelming effects on people's lives," says Kearns. (Page 38)

GALWAY KINNELL is the author of several poetry collections including *Body Rags, The Book of Nightmares,* and *The Avenue Bearing the Initial of Christ into the New World.* He has also published a novel, a book of interviews, and has translated European poetry. His honors include Guggenheim and Rockefeller awards and the Pulitzer Prize. He currently lives in New York City, where he teaches and writes. (Page 186)

MAXINE KUMIN won the Pulitzer Prize in 1973 for *Up Country: New and Selected Poems.* In 1981 and 1982 she served as Poetry Consultant for the Library of Congress. Her most recent poetry collection, *Our Ground Time Here Will Be Brief,* was published in 1982. She and her husband raise horses in New Hampshire. (Page 97)

DAVID LEHMAN reviews books for *Newsweek* and the *Washington Post.* He is the editor of two collections of literary criticism and recently completed a poetry book called *Plato's Retreat.* (Page 135)

ALAN LIGHTMAN was born in 1948 and grew up in Memphis, Tennessee. He was educated at Princeton and the California Institute of Technology, where he was a National Science Foundation Fellow. Since 1976 he has taught astronomy and physics at Harvard, and since 1979 has been a staff member of the Smithsonian Astrophysical Observatory in Cambridge. He has published over fifty articles in scientific journals and two textbooks. Lightman's essays and commentary on science have appeared in *Smithsonian, Science 84, The Boston Globe* and *The New York Times*. Since 1982 he has been a columnist for *Science 84*. Lightman's poetry won the Rhysling Award in 1983. He has consulted for *Sesame Street* and is science advisor for the PBS series *Smithsonian World*. Lightman is the author of *Time Travel and Papa Joe's Pipe: Essays on the Human Side of Science*. He lives in Concord, Massachusetts, with his wife and daughter. (Pages 46, 170)

KATHERINE AUCHINCLOSS LORR is a poet who currently lives in Maryland. Her poems have been widely published regionally and her work has appeared in *Science 82* and *Science 84*. (Pages 60, 111)

CYNTHIA MACDONALD was born in New York City and educated at Bennington College and Johns Hopkins University. A former opera singer, she is now the director of the writing program at the University of Houston. Her books include (W)holes, Transplants and Amputations. (Pages 199, 200)

JOSEPH MATUZAK'S poetry has been published in a number of literary magazines and in anthologies. He has received writing fellowships and awards at the University of Michigan. Currently he works for the Flint Red Cross and also part-time in a computer software store. "I think I use science in my poems because facts are the haven of metaphor and symbol, and science is the home of many wonderful facts," he says. "I agree with Bronowski's idea that both science and art involve a process of reduction and then synthesis, though I trust that most scientists, being empirical, are less casual with the truth than poets." (Page 133)

DAVID MCALEAVEY was born in Wichita, Kansas, in 1946. Sputnik went off, he points out, when he was in sixth grade. He used to have a chemistry lab in his basement and thought he would become a psychiatrist. But he became interested in poetry in high school and grew more serious about writing while attending Cornell University. The poetry that most intrigued him was the radical work that followed upon Eliot and Pound. "The point is that the Pound wing of American poetry is pro-science: no slither." He grew increasingly interested in creating free-verse with rhythmical and musical patterns which might be capable of locking unique meanings into place in mechanically perfected poems. "One poem, only one meaning, the same for everybody. If you can't get it, you can't get at it; but there's only one way to get it. Obscurity as a version of high science." McAleavey teaches writing and literature at George Washington University and has published several volumes of poems. He lives in Arlington, Virginia, with his wife and two young children. (Pages 66, 114)

HEATHER MCHUGH, born to a marine biologist and English literature lover (one of each), has always lived by salt water and the music of language. Half the year she teaches creative writing in graduate programs and half the year she lives on fish and potatoes in Maine. A graduate of Radcliffe College, 1969, she is the

217

author of two collections of poetry and a collection of translations. She's convinced of the felt and family relation between artist and scientist: both know that (as Ginsberg put it) "mind is shapely;" both look for shapeliness in the world. (Pages 136, 149)

PETER MEINKE is director of the Writing Workshop at Eckerd College, St. Petersburg, Florida. He has published two collections of poems, *The Night Train and the Golden Bird* and *Trying to Surprise God*. His poem "Mendel's Law" was the first poem published in *Science 81* magazine. (Pages 7, 129)

ROBERT MEZEY is a poet who teaches at Pomona College and Claremont Graduate School. He has taught and given readings throughout the country and published a dozen volumes of poetry. He has also translated collections of Hebrew and Spanish poetry and edited a major anthology called *Naked Poetry*. He is untrained in science but feels the influence of Loren Eiseley, Louis Thomas, and scientist-poet Miroslav Holub. He has won several awards, including a Guggenheim Fellowship. (Pages 107, 155)

RICHARD MOORE lives in Belmont, Massachusetts. His book, *The Education of a Mouse*, was published last year. His poems have appeared widely in literary magazines, and occasionally in scientific publications. (Page 207)

ROBERT MORGAN was born in 1944 in Hendersonville, North Carolina. He teaches in the Department of English at Cornell University and has published seven collections of poetry. He has published widely in national magazines and anthologies. Morgan has won three National Endowment for the Arts fellowships and several other national prizes. He lives in Ithaca, New York, with his wife and three children. (Pages 48, 65, 83)

G.F. MONTGOMERY was born in 1921 and was educated as an electrical engineer. For many years a government employee, he is now in private practice. Poetry, he feels, is either a hobby or a disease; he is unable to decide. (Page 6)

HOWARD NEMEROV has won a Pulitzer Prize and a National Book Award, both for his collection *Collected Poems*, published in 1978. Born in New York City in 1920, he has also written successful novels. He now lives in St. Louis and teaches at Washington University. (Page 167)

JOHN FREDERICK NIMS, who lives in Chicago, is the editor of *Poetry* magazine. His most recent poetry collections are *The Kiss: A Jambalaya* and *Selected Poems*, both published in 1982. (Page 19)

J. ROBERT OPPENHEIMER, born in New York City in 1904 and died in 1967, was a theoretical physicist and director of the Los Alamos Laboratory during the development of the atomic bomb. He also wrote poetry. (Page 121)

218

DAVID PALMER is the Library Director of the University of Michigan at Flint. He was born in Detroit, Michigan, and now lives in Flint with his wife Charlene, also a poet, and four children. He is a former editor of *The Beloit Poetry Journal* and has published a book called *Quickly, Over the Wall; Poems and Paintings*. His poetry and reviews have appeared in national and local journals, and he has lectured on various aspects of poetry and literature. "I am interested in ideas: philosophy, science, theology, the fine arts, ethics, depth psychology—subjects concerned with the meaning of things. It follows that I like poetry of ideas; poems which have something to say and say it clearly," says Palmer. "The current fashion in poetry of intense preoccupation with polished surfaces and technical acrobatics (the poem as crossword puzzle) I find quite boring." (Page 202)

ANSELM PARLATORE holds a medical degree and has worked in biochemical research and physical anthropology. He has also worked as the editor of two literary magazines, *Granite* and *Bluefish*, and published three books of poetry. He says, "I really don't write poems about science but rather use the language of some of the sciences in my poems because it is the language that I have used or heard during almost every day of my life having spent the greater part of my life in laboratories or in scientific education." (Pages 68, 72, 86, 108, 192)

LINDA PASTAN grew up in New York, graduated from Radcliffe College, and received an M.A. from Brandeis University. She has published five books of poetry; the most recent was a 1982 nominee for the American Book Award. She has received grants from the National Endowment for the Arts and from the Maryland Arts Council, and she is on the staff of the Bread Loaf Writer's Conference. She is married to a physicist, which may account for some of her interest in science. They have three children. (Page 99)

ANNE S. PERLMAN'S first career was as a newspaper reporter. She did not begin writing poetry until she was forty-six years old. A fourth-generation Californian, she received her M.A. degree from San Francisco State University in 1972 and has become an increasingly active voice in the regional poetry community. In 1982 her first book, *Sorting It Out,* was published. Her work often deals with discovery in science and technology. "Perhaps it is the dichotomy inherent in the consequences of science—the power to heal; the power to wreck—that fills me with awe," she says. "This terrible counterpoint distills itself into a poem." Perlman is the mother of three grown children and lives with her husband in San Francisco. (Pages 32, 119, 168)

JONATHAN V. POST writes poems and also writes computer software for the unmanned spacecraft *Galileo,* due for launch in 1988. He has published about one hundred poems, many of them on scientific themes. His prose has appeared in *Scientific American* and *Omni.* (Page 45)

BIN RAMKE studied with R.L. Moore at the University of Texas in a National Science Foundation Summer Program in 1963. His first book, *The Difference Between Night and Day,* won the 1977 Yale Younger Poets Award. He has since published two more poetry collections and teaches English at Columbus College, Georgia. (Page 142)

DANNY RENDLEMAN teaches creative and developmental writing at the University of Michigan at Flint. He has published three books of poetry, including *Signals to the Blind.* He is the editor of *The Journal of English Teaching Techniques.* "Any subject matter is inspiring if it contains the tension of metaphor—of the particular and the universal. It's all a question of wonder and looking closely," he says. (Page 189)

ADRIENNE RICH, born in Baltimore, Maryland, in 1929, is a feminist poet and writer. She won the National Book Award for *Diving Into the Wreck* in 1974. (Page 169)

DAVID ST. JOHN is the author of two collections of poetry, *Hush* and *The Shore.* He teaches in the Writing Seminars of the Johns Hopkins University and is the poetry editor of *The Antioch Review.* (Pages 145, 146)

ROBERT SARGENT was born and educated in the South, but came to Washington as a naval officer in World War II, and remained after the war as a civil servant. He began writing poetry in the 1950s and has been published in many magazines and anthologies. He has published three books of poetry, the most recent of which is *Aspects of a Southern Story,* 1983. He is the former president of the Washington Writers' Publishing House, and on the board of The Word Works, both small presses devoted to poetry. (Page 22)

TIMOTHY SHEEHAN was born in Little Falls, Minnesota, but was raised in La Mesa, California. He was educated at San Diego State University and the University of California at Irvine, where he earned a Masters of Fine Arts. He currently lives in Santa Cruz, California, and has recently become a pilot. "I enjoy the lay study of astronomy and astrophysics, and find the speculations of the scientists engaged in these fields to be incredibly rich and nourishing," he says. (Pages 117, 148)

JUDITH SKILLMAN lives in Bellevue, Washington, with her husband and two children. She earned an M.A. in English literature from the University of Maryland and teaches at Bellevue Community College. Her poems have appeared in several magazines, including *Science 82* and *Science 84.* (Page 137)

MYRA SKLAREW is a poet and writer of short stories who now teaches in the department of literature at American University. She has published four books of poetry; the most recent is *The Science of Goodbyes,* 1983. She was trained as a biologist at Tufts, The Johns Hopkins University and Cold Spring Harbor

Biological Institute, where she worked on bacterial genetics and bacterial viruses. She is currently writing a novel. (Pages 55, 154)

GARY SNYDER was born in 1930 in San Francisco, California. He was educated at Reed College, Berkeley, the University of Indiana and a Zen monastery in Kyoto. In addition to writing poetry, he has worked as a logger, a seaman and a forest ranger. His books include *Turtle Island* and *The Back Country*. (Pages 116, 179)

WILLIAM STAFFORD is the author of more than a dozen collections of poems. His most recent book, *A Glass Face in the Rain*, was published in 1982. He has been honored by a Guggenheim Fellowship for Creative Writing. The former Library of Congress poetry consultant currently teaches English at Lewis and Clark College. (Pages 40, 93)

GEORGE STARBUCK lives with his wife in Milford, New Hampshire. He teaches writing and literature at Boston University. A recent selection of his poetry, *The Argot Merchant Disaster,* includes poems that discuss environmental problems, the creation of the universe and the threat of thermonuclear war. As an undergraduate, Starbuck studied mathematics, until he realized "most great mathematicians do their most serious work by the time they hit twenty." (Pages 23, 113, 195)

WALLACE STEVENS, 1879–1955, an American poet and lawyer, believed that "Poetry increases our feeling for reality and the search for reality is as momentous as the search for God." (Page 91)

ARTHUR STEWART is a field biologist who teaches at the University of Oklahoma and also writes poetry. (Page 54)

MAY SWENSON lives in Sea Cliff, New York, and has published poems in many major national magazines, including *The New Yorker, The Atlantic* and *Science 84*. Her recent book *New and Selected Things Taking Place* includes more than two dozen poems on the topic of science. (Pages 30, 36, 39)

DAVID WAGONER lives in Seattle, Washington, where he teaches at the University of Washington. He is also the editor of *Poetry Northwest*. He has published several books of poetry, the most recent of which is *First Light*. (Pages 8, 174)

DIANE WAKOSKI was born and raised in California, attended Berkeley, and currently lives in East Lansing, Michigan, where she teaches writing at the University of Michigan. She has published more than twenty collections of poetry, including *Collected Greed: Parts I–XIII*, which was published in April 1984. Her poetry has been widely honored with prizes and fellowships; she has recently returned from Yugoslavia, where she was working on a writer's Fulbright. (Page 94)

WALT WHITMAN, 1819–1892, was one of two great nineteenth-century American poets. His collection *Leaves of Grass* is counted among the seminal works of American literature. (Page 164)

RICHARD WILBUR, born in New York City in 1921, is one of the most highly praised poets and translators of his generation. His book, *Things of This World,* 1956, won the National Book Award and the Pulitzer Prize. His verse translations of two Moliére plays enjoyed successful Broadway runs. (Pages 3, 193)

GARY YOUNG is the editor of Greenhouse Review Press in Santa Cruz, California, where he publishes limited editions of poetry books. He has held a fellowship from the National Endowment for the Arts and co-hosts *The Poetry Show* on regional radio. Three collections of his poems have appeared, including his newest *Living In The Durable World,* 1984. "Like many children coming of age in the 1950s, I was schooled in preparation to combat the threat of Sputnik, and the emphasis of my early education was in mathematics and science," says Young. "I only abandoned the idea of a scientific career when I started college and decided poetry was more exact than science in describing and understanding the world." (Pages 77, 115, 118)

ACKNOWLEDGMENTS

Grateful acknowledgment is made to the following for permission to reprint material copyrighted or controlled by them:

Harcourt Brace Jovanovich, Inc., for "Epistemology" from *Ceremony and Other Poems* by Richard Wilbur, copyright © 1950, 1978 by Richard Wilbur. Reprinted by permission of the publisher and the author.

Field Translation Series, #3, Oberlin, Ohio, for "Evening In A Lab" from *Sagittal Section* by Miroslav Holub, translated by Stuart Friebert and Dana Hábová, copyright © 1980 by Oberlin College. Reprinted by permission of the publisher.

Penguin Books Ltd., for "Zito the Magician" from *Miroslav Holub: Selected Poems,* translated by Ian Milner and George Theiner (Penguin Modern Poets 1967), copyright © 1967 by Miroslav Holub, translation copyright © 1967 by Penguin Books Ltd. Reprinted by permission of the publisher.

Science 84 magazine, for "Graham Bell and the Photophone" by G.F. Montgomery, copyright © 1984. Reprinted by permission of the publisher and the author.

Science 82 magazine, for "Hermann Ludwig Ferdinand Von Helmholtz" by Peter Meinke, copyright © 1982. Reprinted by permission of the publisher and and author. This poem was originally published in *Trying to Surprise God* (1981) by Peter Meinke, University of Pittsburgh Press.

Little, Brown and Company, for "My Physics Teacher" from *Landfall: Poems by David Wagoner,* copyright © 1981 by David Wagoner. Reprinted by permission of the publisher and the author.

Isaac Asimov's Science Fiction Magazine, for "Marie Curie Contemplating the Role of Women Scientists in the Glow of a Beaker" by Robert Frazier, copyright © 1983 by Robert Frazier. Reprinted by permission of the author.

Science 84 magazine, for "Letter from Caroline Herschel (1750–1840)" by Siv Cedering, copyright © 1984 by Siv Cedering. Reprinted by permission of the publisher and the author.

Siv Cedering, for "Letter from Nicholas Copernicus" by Siv Cedering, originally published in *Skywriting* in 1976. Also for "Letter from Johannes Kepler" by Siv Cedering, originally published by *Confrontation* in 1976. These poems are forthcoming in the collection *The Floating World* (fall 1984) by Siv Cedering, University of Pittsburgh Press.

Louisiana State University Press for "How Copernicus Stopped the Sun" and "How Einstein Started It Up Again" from *The First Man on the Sun* by R.H.W. Dillard, copyright © 1983. Reprinted by permission of the publisher.

Science 83 magazine, for "The Island of Geological Time" by Laura Fargas, copyright © 1983 by Laura Fargas. Reprinted by permission of the publisher and the author.

Vegetable Box, for "Brevard Fault" by Robert Morgan, copyright © 1978 by Robert Morgan. Reprinted by permission of the author.

Science 84 magazine, for "Time's Time Again" by A.R. Ammons, copyright © 1984 by A.R. Ammons. Reprinted by permission of the publisher and the author.

Random House, Inc., for "Unpredictable But Providential" from *W.H. Auden: Collected Poems,* edited by Edward Mendelson, copyright © 1969, 1973, 1974 by The Estate of W.H. Auden. Reprinted by permission of the publisher.

William Morrow & Company, Inc., for "Ice Dragons" from *Lady Faustus* by Diane Ackerman, copyright © 1983 by Diane Ackerman. Reprinted by permission of the publisher and the author.

Science 84 magazine, for "Fossils" by Arthur Stewart, copyright © 1984 by Arthur Stewart. Reprinted by permission of the publisher and the author.

Science 82 magazine, for "The Origin of Species" by Myra Sklarew, copyright © 1983 by Myra Sklarew. Reprinted by permission of the publisher and the author.

Random House, Inc., for "Progress?" from *W.H. Auden: Collected Poems,* edited by Edward Mendolson, copyright © 1969, 1973, 1974 by The Estate of W.H. Auden. Reprinted by permission of the publisher.

Science 82 magazine, for "Peking Man, Raining" by Katharine Auchincloss Lorr, copyright © 1982 by Katharine Auchincloss Lorr. Reprinted by permission of the publisher and the author.

Science 82 magazine, for "Skull of a Neandertal" by Michael Cadmun, copyright © 1982 by Michael Cadmun. Reprinted by permission of the publisher and the author.

Field Translation Series #7, Oberlin, Ohio, for "Homization" from *Interferon, or On Theater* by Miroslav Holub, copyright © 1982 by Oberlin College, translated by David Young and Dana Hábová. Reprinted by permission of the publisher.

Science 83 magazine, for "Two Sonnets" by Helen Ehrlich, copyright © 1983 by Helen Erhlich. Reprinted by permission of the publisher and the author.

Science 84 magazine, for "Jutaculla Rock" by Robert Morgan, copyright © 1984 by Robert Morgan. Reprinted by permission of the publisher and the author.

Random House, Inc., for "The Question" from *W.H. Auden: Collected Poems,* edited by Edward Mendelson, copyright © 1969, 1973, 1974 by The Estate of W.H. Auden. Reprinted by permission of the publisher.

Robert Frazier, for permission to reprint "The Supremacy of Bacteria" by Robert Frazier, copyright © 1980 by Robert Frazier. Reprinted by permission of the author.

Ithaca House, for "Cancer Research" from *Hybrid Inoculum* by Anselm Parla-

Science 84 magazine, for "Waiting for *E. gularis*" by Linda Pastan, copyright © 1984 by Linda Pastan. Reprinted by permission of the publisher and the author.

Alfred A. Knopf, Inc., for "The Cormorant in Its Element" and "Camouflage" from *The Kingfisher* by Amy Clampitt, copyright © 1982 by Amy Clampitt. Reprinted by permission of the publisher and the author. "The Cormorant in Its Element" was originally published in *The Atlantic Monthly*. "Camouflage" was originally published in *The New Yorker*.

Field Translation Series #3, Oberlin, Ohio, for "Teaching about Anthropods" and "Brief Reflection on the Insect" from *Sagittal Section* by Miroslav Holub, translation by Stuart Friedbert and Dana Hábová, copyright © 1980 by Oberlin College. Reprinted by permission of the publisher.

Robert Mezey, for "In This Life" by Robert Mezey, copyright © 1974 by Robert Mezey. This poem was originally published by Kayak Press. Reprinted by permission of the author.

Ithaca House, for "Accommodation" from *Hybrid Inoculum* by Anselm Parlatore, copyright © 1974 by Ithaca House. Reprinted by permission of the publisher and the author.

Times Books/*The New York Times* Book Co. Inc., for "The Spider" and "Winter Sign" from *The Star Thrower* by Loren Eiseley, copyright © 1978 by The Estate of Loren Eiseley. Reprinted by permission of the publisher.

Science 84 magazine, for "The Beekeeper's Dream" by Katharine Auchincloss Lorr, copyright © 1984 by Katharine Auchincloss Lorr. Reprinted by permission of the publisher and the author.

The Christian Science Monitor, for "Amphibian" by Amy Clampitt, copyright © 1983 by the Christian Science Publishing Society. All rights reserved. Reprinted by permission of the publisher and the author.

Little, Brown and Company, for "Pit Viper" from *The Argot Merchant Disaster: Poems New and Selected* by George Starbuck. Reprinted by permission of the publisher, in association with the Atlantic Monthly Press, and the author. This poem was originally published in *The New Yorker*.

Jazz Press, for "Equinox" from *Hands* by Gary Young, copyright © 1980 by Gary Young. Reprinted by permission of the author.

New Directions, for "Once Only" from *The Back Country* by Gary Snyder, copyright © 1968 by Gary Snyder. Reprinted by permission of the publisher and the author.

Greenhouse Review Press, for "At Camino" from *YES* by Timothy Sheehan, copyright © 1979 by Timothy Sheehan. Reprinted by permission of the publisher and the author.

The Bieler Press, for "Tornado Watch, Bloomington, Indiana" from *In the Durable World* by Gary Young, copyright © 1984 by Gary Young. Reprinted by permission of the publisher and the author.

Carnegie-Mellon University Press, for "At Liberty" from *Sorting It Out* by Anne S. Perlman, copyright © 1982 by Anne S. Perlman. Reprinted by permission of the author.

Science 83 magazine, for "Hieroglyphic" by Myra Sklarew, copyright © 1983 by Myra Sklarew. Reprinted by permission of the publisher and the author.

Robert Mezey, for "Song" by Robert Mezey, copyright © 1978 by Robert Mezey. This poem was originally published by Humble Hills Press.

Field Translation Series #3, Oberlin, Ohio, for "Poem Technology" and "Newborn Baby" from *Sagittal Section* by Miroslav Holub, translated by Stuart Friebert and Dana Hábová, copyright © 1980 by Oberlin College. Reprinted by permission of the publisher.

Science 84 magazine, for "Doorman" by Martin Galvin, copyright © 1984 by Martin Galvin. Reprinted by permission of the publisher and the author.

Liveright Publishing Corporation for "La Guerre" from *Tulips and Chimneys* by e.e. cummings, copyright © 1923, 1925 and renewed 1951 and 1953 by e.e. cummings; copyright © 1973, 1976 by the Trustees for the e.e. cummings Trust. Copyright © 1973 and 1976 by George James Firmage. Reprinted by permission of the publisher.

Howard Nemerov, for "The Weather of the World" from *Poems by Howard Nemerov* (University of Chicago Press), copyright © 1978 by Howard Nemerov. Reprinted by permission of the author.

Carnegie-Mellon University Press, for "The Specialist" from *Sorting It Out* by Anne S. Perlman, copyright © 1982 by Anne S. Perlman. Reprinted by permission of the author.

W.W. Norton and Co., for "Artificial Intelligence" from *Snapshots of a Daughter-in-Law* by Adrienne Rich, copyright © 1961, 1978 by Adrienne Rich. Reprinted by permission of the publisher.

Science 82 magazine, for "In Computers" by Alan P. Lightman, copyright © 1982 by Alan P. Lightman. Reprinted by permission of the publisher and the author.

Random House, Inc., for "Moon Landing" from *W.H. Auden: Collected Poems*, edited by Edward Mendelson, copyright © 1969, 1973, 1974 by The Estate of W.H. Auden. Reprinted by permission of the publisher.

Science 84 magazine, for "Koko" by Ann Downer, copyright © 1984 by Ann Downer. Reprinted by permission of the publisher and the author.

Little, Brown and Company, for "In Distress" from *Landfall: Poems by David Wagoner,* copyright © 1981 by David Wagoner. Reprinted by permission of the publisher and the author. "In Distress" first appeared in *Poetry Northwest.*

New Directions, for "Toward Climax" from *Turtle Island* by Gary Snyder, copyright © 1974 by Gary Snyder. Reprinted by permission of the publisher and the author.

Alfred A. Knopf, Inc., for "Berceuse" from *The Kingfisher* by Amy Clampitt, copyright © 1982 by Amy Clampitt. "Berceuse" was originally published in *The New York Review of Books.*

Science 84 magazine, for "Of How Scientists Are Often Ahead of Others in Thinking, While the Average Man Lags Behind; and How the Economist (Who Can Only Follow in the Footsteps of the Average Man Looking for Clues to the